The Science of Structure: Synergetics

Hermann Haken

THE SCIENCE OF STRUCTURE: SYNERGETICS

Translated by Fred Bradley

VAN NOSTRAND REINHOLD COMPANY
NEW YORK CINCINNATI TORONTO LONDON MELBOURNE

English translation by Fred Bradley

Library of Congress Catalog Card Number 83-10198
ISBN 0-442-23703-0

Printed in the United States

Published by Van Nostrand Reinhold Company
Inc.
135 West 50th Street
New York, New York 10020

Van Nostrand Reinhold Company Limited
Molly Millars Lane
Wokingham, Berkshire RG11 2PY, England

Van Nostrand Reinhold
480 La Trobe Street
Melbourne, Victoria 3000, Australia

Macmillan of Canada
Division of Gage Publishing Limited
164 Commander Boulevard
Agincourt, Ontario M1S 3C7, Canada

16 15 14 13 12 11 10 9 8 7 6 5
4 3 2 1
Library of Congress Cataloging in Publication Data
Haken, H.
The science of structure, synergetics.

Translation of: Erfolgsgeheimnisse der Natur.
Includes bibliographical references and index.
1. Biology—Philosophy. 2. Natural history—
Philosophy. I. Title.
QH331.H3413 1984 574'.01 83-10198
ISBN 0-442-23703-0

CONTENTS

Preface

Nature, especially the animal and vegetable kingdoms, is an inexhaustible source of surprise to us through the abundance of its forms and the delicacy of its structures, in which the individual parts cooperate most ingeniously. Earlier generations regarded these structures as God-given. Today scientific interest is increasingly turning to the question of how these structures originate and what forces are at work here. Whereas until recently it looked as if the autogenic origin of structures contradicted the principles of physics, this book represents a turning point of scientific thought. Our point of departure is the realization that even in inanimate matter novel, well-ordered structures grow from chaos and can be maintained with a constant supply of energy. This book provides most instructive evidence of that discovery in physics and chemistry, such as the orderly arrangement in the laser beam, honeycomb patterns in liquids, and chemical spiral waves. As we shall see from these examples, the formation of structures is based on laws of universal validity. This realization will enable us to deal with even more complex problems, such as the question of how the pattern formation of cells in animals is controlled, how collective behavior patterns of commercial companies can determine economic events, and what are the rules that determine the formation of public opinion in our society.

Almost always in these processes a very large number of individual factors must cooperate in a meaningful manner. We are faced with "complex systems," as they are called. These can be looked at from various angles: the function of the individual components can be investigated, or the system examined as a whole. In the first case we shall start, as in a game, from rules that determine the individual steps of the various parts and thereby eventually produce a "pattern." This is shown convincingly in the book *Das Spiel* (The Game) by Manfred Eigen and Ruthild Winkler.

Synergetics, "the science of cooperation," follows the second path. Here we rarely inquire into the individual elementary rules; we want to discover the general laws according to which structures form. Although all comparisons are odious, we would liken synergetics to a game of chess. We can play this game time and again, following the moves of the individual figures. But we can also ask: what can we say about the final state of a game of chess? Obviously, everybody is familiar with this situation: either the white king or the black king is beaten, or the result is a draw. Although in their totality the individual moves are extremely complex, the final outcome can be represented in a few words. Something similar happens with the structure formations examined here: we ask for the ultimately-forming global patterns. We shall thus recognize the existence of general, higher-ranking cogencies that lead to the new structures, the new patterns.

The knowledge gained in the field of science about collective behavior also has a very decisive bearing upon our personal affairs, whether in the economic or in the social sphere. This book, however, does not offer ready solutions in that area. It is to be hoped the book stimulates thought; it does not provide prescriptions for our own behavior. We shall even formulate and prove the thesis that it is sometimes quite impossible to find clear-cut solutions, which also throws new light on the nature of conflicts and how to deal with them.

The field of synergetics is developing at a frantic pace, which is evident from the growing number of international conferences, and as well from the fact that the VW Foundation promotes synergetics in the natural-scientific and technological field by means of a program focused on those disciplines. Springer Publishers is dedicating its Springer Series in Synergetics to it. The aim of this book is to introduce this new field, already so fascinating to the scientists, to the interested layman as well.

We hear so much nowadays about the "debt" science owes us. It seems to me that science and society exist in an inseparable symbiosis. Society is as vital to science as science is to society. Every bridge built between them is therefore important. To repay his debt is not an easy matter for the scientist. The will is there, on the whole, but the language of science—especially when it makes use of mathematics—has become so divorced from everyday language that translation has become very difficult. I would nevertheless claim that a process, whether it occurs in the natural sciences or, for instance, in economics, will often be fully understood by the scientists only when it can also be explained in everyday language without any formulae. The very need to make himself clear to the layman provides the scientist with new insight into major interconnections.

I hope that my description of this new scientific discipline will stimulate and create suggestions about how to utilize the secrets of nature's successes for your own benefit and for that of the whole of mankind.

My thanks are due to my wife for her critical scrutiny of the manuscript and her valuable suggestions for improvement, and to Mrs. Ursula Funke for her quick and perfect production of the typescript. Her untiring enthusiasm assisted me greatly in bringing this book to a successful conclusion.

I am also grateful to the staff of Deutsche Verlags-Anstalt, especially Dr. Lebe and Mrs. Locke, for their valuable cooperation.

HERMANN HAKEN

Stuttgart, Spring 1981

Chapter 1

Introduction and Survey

Why This Book Might Interest You

Our world consists of the most diverse objects. Many of them are man-made: houses, automobiles, tools, paintings. But many others are the products of nature. To the scientist this world of objects becomes a world of structures and orders subject to strict laws. When we point our telescopes into the immeasurable depths of the universe, we see the spiral nebulae of which figure 1.1 shows an example. The spiral arms to which the nebula owes its structure, its arrangement, are clearly visible. New, brightly shining suns in unimaginably large numbers form in them out of gas. Our earth and sun, too, belong to a spiral nebula, the Milky Way, which on clear nights we can distinctly see in the sky. Our sun is only one of a hundred billion others in it—a number beyond human comprehension. The planet earth, together with the other planets, orbits the sun according to immutable laws.

But to find structures we need not look at the universe at all. Our everyday surroundings provide us with countless examples. One of them is the snow crystal in its regular structure (fig. 1.2). Animate nature never fails to surprise us with the abundance of its forms, which may be quite bizarre. Figure 1.3 is the magnified picture of the eye of a tropical fly, which is mounted on a stalk that protrudes from the head. The honeycomb structure of the eye is fascinating in its regularity. At

FIGURE 1.1 A spiral nebula.
FIGURE 1.2 A snow crystal.
FIGURE 1.3 Eyestalk of a tropical fly (Diopsis thoracica). The hexagonal structure is a conspicuous feature.

the same time the whole "construction" seems to make excellent sense, as it provides a perfect view all around. We are often delighted by the harmony of the forms of animals and plants we encounter in incredible variety. We frequently find the structure of the creatures most practical; but when we look at the gorgeous flowers of many plants, nature strikes us also as playful and capricious.

But it is not only immobile structures that we admire; we also enjoy movement sequences in their orderliness, such as the cantering of a horse or the gracefulness of a dance. In the human community we find structures on a higher plane. Society is ordered in certain forms of state that differ radically from each other. In the purely intellectual world, too, we encounter structures—in language, in music, and in the world of science. We are thus constantly faced, from inanimate to animate nature and to the world of the intellect, with structures; and we have become so used to them that we are no longer conscious of the miracle of their existence.

In the past, people regarded structures as God-given, as the Old Testament story of the creation, for instance, makes clear. Scientists, too, concentrated exclusively on the question of how structures are arranged, not how they originate. Only recently has the interest of researchers become increasingly focused on the latter question. Unless we want to posit a supernatural power—i.e., claim a new act of creation each time we want to explain such structures—science is faced with the task of explaining how structures come about spontaneously, or, in other words, how they organize themselves.

In Pursuit of a Unified Picture of the Universe

In view of the abundance of all these structures, when we ask ourselves how they originated, this is at first sight an undertaking whose end we cannot foresee. Breaking down the structures into their components occupied the time of many generations of scientists and is still continuing, so does not the unlocking of the secret of their origin require much more work and effort still? Indeed, if the origin of each single structure were subject to highly specific laws applicable to it alone, it would not be enough to write just one book; the knowledge would have to be recorded in a truly enormous library.

Here we are confronted with an idea that is the thread running through all the fabric of science: the collection of facts on the one hand is balanced by the urge to develop a unified picture of the world on the other. We are familiar with these efforts especially in the natural sciences, in physics, chemistry, and biology, and also in philosophy. The search for fundamental laws of physics is not new to us. Sir Isaac Newton's (1642–1727) laws of motion and his law of gravity explain the motion of the planets round the sun, a motion that the ancients were unable to understand on a common premise. James Clerk Maxwell (1831–1879) enabled us to grasp that light is nothing but an electromagnetic vibration, exactly like a radio wave. Albert Einstein (1879–1955) succeeded in linking gravity with space and time. In chemistry, Dmitri I. Mendeleev (1834–1907) was the first to bring order into the abundance of chemical substances by introducing the periodic system of elements. Modern atomic physics was able to derive this system from the basic laws of the atomic structure. In biology, Mendel's laws show how characteristics are inherited, for instance when flowers of different colors are cross-bred. In our own days the discovery of the chemical basis of heredity in the form of giant biological molecules, the so-called DNA (deoxyribonucleic acid), has completed the picture.

As these examples show, and we could list many dozens, universal fundamental laws governing the work of nature are being discovered all the time. Whereas on the one hand the most varied phenomena can be reduced to a few basic laws, on the other hand research is continuously uncovering new facts of even more complex reactions, and we are often close to being overwhelmed by the vast quantities of such material. There is a neverending race in science between the deluge of new facts and their categorization, discovery of their significance and their conformity with general laws.

Dissection or Construction

How can we hope to understand structures or processes within them? It is a popular and often successful approach to dissect objects being investigated into progressively smaller components: the physicist splits up the crystal (which shall be discussed in greater detail in Chapter 3) into its constituents, the atoms, and breaks down the atoms into ever smaller particles, the atomic nuclei and the electrons. An important branch of modern physics research deals with even "more elementary" particles, the quarks and the gluons, neither of which may yet be the ultimate building bricks of matter. The biologist teases cells from tissue and dissects them into their components, such as cell membranes and cell nuclei, which in turn are dismantled into their components in the form of biomolecules. We could add countless other examples from the most varied branches of science. Indeed, even science itself is split up into its various branches—mathematics, physics, chemistry, all the way to sociology and psychology.

When the researcher uses this method, however, his experience may be quite similar to that of the little boy given a toy automobile: soon the boy will want to know why the automobile runs and he will dismantle it into various parts, generally something he will manage without much difficulty. But often we will find him sitting in tears in front of the bits and pieces because he still does not know what enables the automobile to run; nor is he able to put them together again to form a whole that makes any sense. He thus learns early the meaning of the maxim that the whole is more than the sum of its parts; or, as Goethe expressed it, "I am now holding the parts in my hand; alas, the mental link is still missing." Applied to the most varied sciences, this means that even after we have discovered the nature of structures we still need to understand how the individual components cooperate.

As we shall see, the question of how the structures originate is closely related to this and has led to the establishment of synergetics, which deals with this complex of problems. The term, like so many other scientific terms, is derived from the Greek, and means "the science of cooperation." We hope the concept will allow us to discover, in spite of the abundance of the most varied structures found in nature, whether we can identify unified fundamental laws that shed light on how structures are created.

This obviously sounds very obscure, very abstract. I must admit that the only precise answer is based on a mathematical theory; I have established its validity in a wide range of applications. On the other hand, the very profusion of the various examples at our disposal enables us to describe the basic reactions quite lucidly. We can resort to simple examples, for instance in the field of mechanics. This does not mean that I want to design a mechanistic picture of the world. But our language derives many words from mechanistic terms, such as "balance." Originally this conjures up the picture of a pair of scales with two equal weights. The scales do not move; they are "in balance." But when we speak of mental balance, nobody would ever think of claiming that we interpret our intellect in mechanistic terms. The reader of this book is well advised to remember this example, because I do not want to deal only with structures in the material world but also in the world of ideas, such as economic or cultural developments.

Do Biological Structures Contradict Fundamental Natural Laws?

Physics claims to be the fundamental natural science par excellence. It deals with matter, and as everything consists of matter, all matter must obey the laws of physics. This conviction, among the biologists for instance, was by no means always held. The vitalists insisted that living creatures had their own, quite specific life force. After scientists had succeeded in tracing chemical reactions basically to physical reactions (such as chemical bonds or the atomic structure), hardly any doubt persisted that it should theoretically be possible to explain fundamental biological reactions in terms of physical ones. We have, by the way, left ourselves a loophole in this claim with the qualification "theoretically." We shall explain presently that this concerns a whole complex of questions.

Let us, for the time being, discuss the much too simplistic statement that the laws of physics also apply to biology. Those who took seriously the claim that biology could be reduced to physics very quickly became entangled in contradictions only a few years ago. If a physicist had been asked whether the origin of life conformed to the fundamental laws of physics, he would in all honesty have had to deny it. Why? According to the fundamental laws of physics, more strictly according to the laws of thermodynamics, chaos in the universe ought to be progressively increasing. All regulated function sequences ought to cease, all orders to disintegrate.

The only way out of this dilemma that many leading physicists saw was to regard the creation of orderly states in nature as a phenomenon of enormous fluctuation that, according to the rules of the probability theory, should moreover be of random improbability. This is a truly absurd idea that, however, seemed to be the only acceptable one within the framework of so-called statistical physics.

I shall explain in Chapter 2 why the physicists thought that disorder should progressively increase. It will become obvious that physics has left itself a vital loophole for the formation of structures (e.g., of the crystals). But as we shall learn these structures are inanimate. Had physics maneuvered itself into a dead end by claiming that biological reactions are based on physical laws, but that the origin of life contradicts the fundamental laws of physics? A lucky coincidence helped us break out of this vicious circle: we discovered that physics itself provided a perfect precedent for events in which a certain type of animate order is created, even though it strictly conforms to the laws of physics, and in fact would be impossible without them. It is the laser, a new light source that is already widely known. This example will show us that inanimate matter, too, can organize itself to produce apparently meaningful events.

We shall encounter very remarkable conformities with natural laws that are common to all phenomena of self-organization (fig. 1.4). We shall realize that the individual components arrange themselves as if prompted by an invisible hand, but conversely it is these individual systems that in turn create this invisible hand by their cooperation. Let us call this invisible hand that organizes everything the order parameter. But again, we seem to be caught in a vicious circle.

The order parameter is created by the cooperation of the individual parts; conversely the order parameter rules the behavior of the individual parts. It is the old question of which came first, the chicken or the egg. (Not a word about the cock.) In terms of synergetics, the order parameter enslaves the individual parts. The order parameter is like a puppeteer who lets the puppets dance, but the puppets in turn have an effect on him, control him. As we shall find out, the principle of enslaving plays a central part in synergetics. We must, however, point out that this involves no value judgment; it merely expresses a causal relation and has nothing to do with enslavement in the ethical sense. Thus the members of a nation, for instance, are enslaved by its language.

When I examined other phenomena, first of physics, then of chemistry, and finally of biology, from the aspect of order parameter and enslavement I found the same evidence time and again. The processes of structure formation somehow advance inevitably in a certain direction, but by no means as predicted by the laws of thermodynamics, by no means even in the direction of ever-increasing disorder. On the contrary, still irregular part systems are also engulfed in the existing state of order, and their behavior is enslaved by it.

This inevitability of the creation of order from chaos is, as we shall

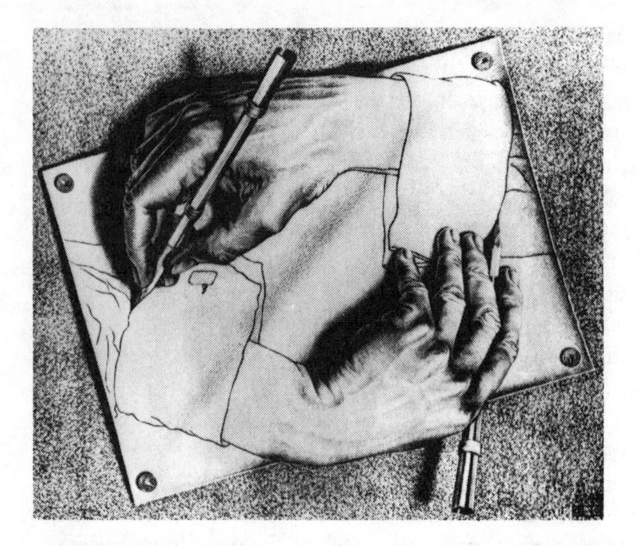

FIGURE 1.4 Escher's picture of two hands that are reciprocally drawing each other demonstrates the problems of self-organization: the order parameter (one hand) causes the behavior of the components (the other hand); its behavior is in turn determined by the behavior of the components.

discover, largely independent of the material substrate on which the reactions occur. In this sense a laser may behave exactly like a cloud formation or an accumulation of cells. We are obviously confronted with a uniform phenomenon. This would suggest that such laws would also be found in the nonmaterial area. One of the examples of this in sociology is that the behavior of entire groups seems suddenly to succumb to a new idea, a fashion perhaps, or to intellectual currents such as a new direction in painting or a new literary style.

As will soon become obvious, these natural laws are the key that enables us to unlock the secrets of nature's success. How does nature manage, for instance, to evolve more and more complex species in the animate world? How do some species find the means to gain increasing ascendancy and to displace others? On the other hand, how is it possible that, despite the most ruthless competition, species can coexist, in fact stabilize each other by this very coexistence? From this new aspect, phenomena regarded in the past as isolated become examples of a unified natural law. What was hitherto puzzling, indeed contradictory, suddenly falls into place. We shall find that it is the collective behavior of many individual units, whether atoms, molecules, cells, animals, or human beings, that indirectly determines their fate through competition on the one hand and cooperation on the other. But they are more often pushed than doing their own pushing.

In this sense, we can regard synergetics as a science of orderly, self-organized, collective behavior subject to general laws. When a science makes pronouncements of great general validity, there are bound to be some important consequences. Synergetics encompasses a great variety of disciplines, such as physics, chemistry, biology, as well as sociology and economics. We therefore expect that the laws discovered and described by synergetics are already represented, more or less hidden, in various disciplines. Synergetics thus creates a new picture out of many isolated facts, much in the nature of a jigsaw puzzle.

A second consequence must not be overlooked. Science has consistently taught us that it is rash to regard laws as universal. Time and time again we have had to accept that natural laws, their validity recognized and confirmed in certain fields, were only an approximation within a larger framework or lost their significance altogether. Thus Newton's mechanics are only an approximation of those of Einstein's relativity theory. Classical mechanics, which describes the motions of macrobodies, had to be replaced by quantum mechanics in the submicroscopic world of the atoms. In this sense synergetics is more comprehensive than the laws of thermodynamics, its range of application considerably wider. On the other hand, synergetics too has its limitations. In explaining these limitations we must distinguish between its aims and what it has so far achieved. It is the aim of synergetics to establish the natural laws on which the self-organization of systems in the most diverse fields of science is based. It has succeeded in finding such general laws especially for the most interesting cases—i.e., for those in which structures are newly created or the macrostates of systems change drastically. But what are macrosystems and what is drastic? Examples can illuminate the terms better than long explantions. I shall present a large number of them in the hope of introducing the reader step by step to the problems of synergetics on the one hand and to its findings on the other.

All vital processes, all the way from those of a single cell through to the coexistence of mankind and nature, are very closely meshed; all parts engage each other like gear wheels, directly or indirectly, which makes the systems always very complex. Rising population and density and advancing technology are increasing the complexity of our environment; consequently the task of understanding the behavior of complex systems is continually expanding. Here synergetics opens up a fundamentally new insight, as I shall have frequent occasion to show. A complex system is like a thick book that we should read from cover to cover in order to know it inside-out and understand its contents fully. But what to do if time is scarce? We can proceed in different ways. For instance, we can read salient passages. Or someone may give us a short précis, which may have been selected from a great variety of aspects. One reader may consider the love affair in the book the most important feature, another perhaps the social environment it describes. A book can also be classified under one or more headings, such as a "historical novel," "nonfiction," "whodunnit," and so on. Other characterizations too are of interest, such as "dud" or "bestseller." Because the human brain (even the collective brain of all scientists together) can accept only a limited information content, we must treat complex systems like an overlong book: we must look for condensed information that is relevant to our purpose.

Even if we were able to collect all the data, it would cloud our judgment more than assist it. We would no longer be able to see the forest for the trees. Hardly another aphorism illuminates the problems of complex systems better. We must not allow ourselves to be bogged down in unimportant detail. We must learn to see and to grasp the general context. We must "reduce the complexity."

As synergetics shows the "relevant information," the overall context is supplied by the order parameters, which will become most significant whenever the macrobehavior of the systems changes. Generally these order parameters are the long-lived magnitudes that enslave the short-lived ones. Numerous examples will confirm this.

If such general laws apply where order is created from chaos or an existing order gradually changes into a new one, a certain automatism is bound to be inherent in these events. When we learn to identify these laws as well in the economic, sociological, or political spheres, we shall find it easier to cope with the difficulties of life. We shall realize, for instance, that an attitude of others directed against us is not based on a conspiracy against us but that the others act, in fact are compelled to act, because of certain collective modes of behavior. Knowledge of these automatisms may even ensure that they work for and not against us. As we can lift the heaviest loads with little effort when we apply the law of the lever, the application of synergetic laws produces great effects with little effort. We can thus make use of the "secret of nature's successes" for our own benefit.

We shall see time and again that the animate world was able to develop as far as it did only because it did not have unlimited resources. It can give its processes only limited time in which to develop, consistent with our experiences in our own lives. It was precisely these external constraints that spurred on development in nature and produced ever new kinds of creatures. I do not consider it a coincidence that technology, civilization, progressed farthest in those countries that do not experience the debilitating heat of perennial summer but have to contend with the biting cold of winter.

As we enter this new field of synergetics we must obviously proceed from simple to more and more complex events. We shall therefore begin with examples from physics and chemistry before tackling questions of economics, sociology, and the theory of science. The method of applying experiences gained from simple examples to more complex ones is not new. In sociology and economics, for instance, models have been developed that are oriented according to those of physics and make extensive use of the physical concept of entropy, which is a measure of disorder.

With the new knowledge gained in physics, however, fresh thinking is also already advancing in other branches of science. Whereas in the past the structure of a given society was regarded as static, in balance, our contemporary way of looking at it has changed completely. Structures form, disappear, compete, cooperate, or combine into larger structures. We have reached a turning point in our thinking, where we move from statics to dynamics.

Before we can deal with all these questions we must come to grips with the basic objection of physics to the formation of structures: the principle of progressive disorder.

Chapter 2

Is Disorder Progressive? The Thermal Death of the World

The One-Way Street of Nature

Physics has the advantage of letting natural events run their course in precisely defined experimental conditions. By noting that the events take place in always the same manner it is able to formulate natural laws of general validity. Some of these laws are evident even in our daily lives. When we heat an iron bar at one end, the temperature in it becomes equalized after some time (fig. 2.1). The reverse event, of an iron bar spontaneously becoming hot at one end and cold at the other, has never been observed. When we bring two vessels together, one filled with gas, the other empty, and remove the partition between them, the gas will expand with a hiss into the empty vessel until both vessels are equally full (fig. 2.2). The reverse event, in which the molecules of gas in a vessel suddenly congregate in one half, has also never been observed. When we apply the brakes to a moving car, the automobile will eventually stop, with the brakes and possibly also the tires becoming hot. On the other hand, heating the brakes and tires has never yet started up an automobile. It is obvious that all these natural events proceed in one direction only. Their reversal is forbidden in nature; this is why they are called irreversible. During the last century, the brilliant Austrian physicist Ludwig Boltzmann (1844–1906) succeeded in providing a first decisive answer to the question of why nat-

FIGURE 2.1 The temperature of an iron bar heated at one end equalizes itself. The result is a warm iron bar.

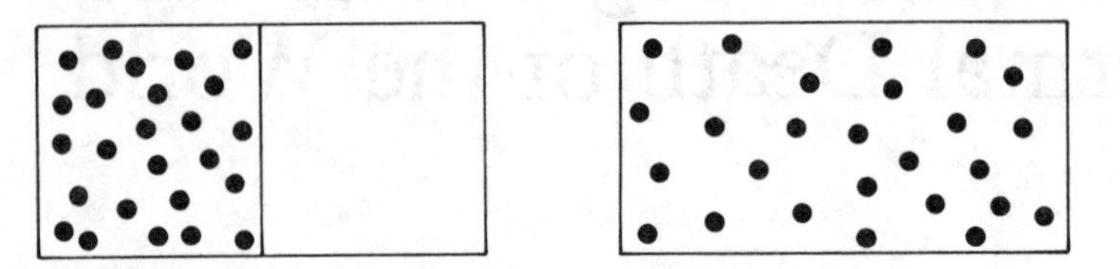

FIGURE 2.2 If the partition between the two vessels on the left is removed, the gas atoms will fill both vessels equally.

ural events proceed in a certain direction; it was that the events proceeded in the direction of progressive disorder.

What Is Disorder?

But how can we define disorder?

The physical concept of disorder is quite closely related to the concept of disorder we know in our daily life. Why, for instance, is a child's room disorderly? Well, because it has not been straightened up or, in other words, the various items (notebooks and schoolbooks, say) are not where they should be (fig. 2.3). The biology book, for instance, is not in its allotted place on the shelf but on the table, on the window sill, on the chair, on the bed, on the floor, or anywhere else. There are so very many possibilities for where it could be. The same applies to a notebook, pen, eraser. If all the items are in their allotted places we have the state of the straightened room—i.e., of order. There is thus exactly one state of order. At the same time, it is obvious that disorder is associated with the large number of the different possibilities where any given object could be. This is precisely why it is so difficult to find the item we are looking for in a disorderly state. It is, we repeat, the larger number of different possibilities of where something could be that constitutes the state of disorder.

FIGURE 2.3 Order and chaos as represented by the artist Escher. In chaos obviously nothing is where it belongs (for instance in the garbage can).

The multitude of the various possibilities is also the measure of disorder in physics. We can see this in the very simple example of the gas. Let us look at the model of a gas that consists of only four molecules, consecutively numbered 1–4, and which we distribute in two boxes.

There is only *one* possibility of putting all four molecules in a given box, for instance in the one on the top left of figure 2.4. But there are six possibilities, as this illustration proves, of distributing the spheres in the boxes two each, differently but equally. In the macrostate, that is expressed superficially: either all the molecules are in one box, or they are half-and-half in each of the two boxes. According to Boltzmann's principle, nature aims at those states that leave the largest number of possibilities open. The term "entropy" used by physicists is, as Boltzmann laid down, determined by the number of these possibilities

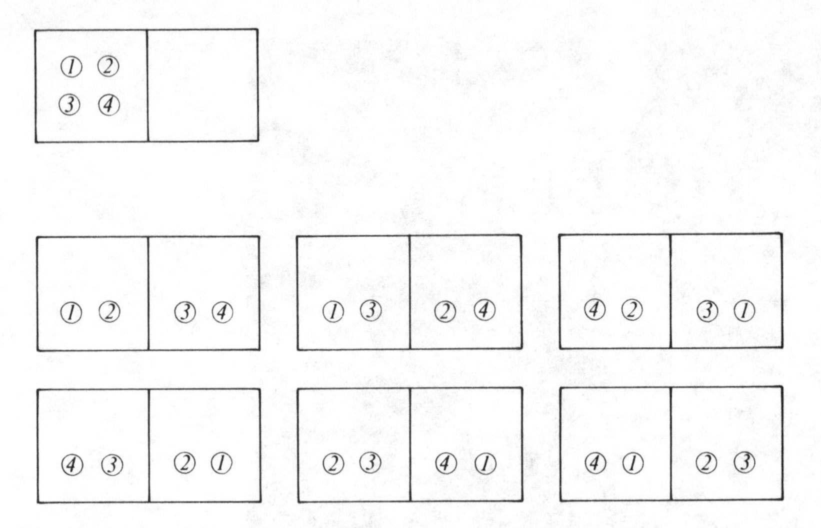

FIGURE 2.4 Illustration of Boltzmann's counting rule for the determination of maximum entropy *Top:* Only one possibility of housing the four balls in a single vessel. *Bottom:* The six possibilities of equally distributing the four balls among the two vessels.

(more accurately, by the logarithm of this number). Nature thus aims at the state of highest entropy.

In our example of the four molecules, the six possibilities of "equal distribution" are contrasted with the single one of "all molecules in one box." In reality the number of gas molecules, for instance in 1cc, is already enormous, and accordingly the number of individual possibilities of equal distribution in both boxes is simply immeasurable. As a result, the chance that nature realizes the state of equal distribution is also extraordinarily great, and all deviations from it are at worst minor fluctuations, for instance a small density fluctuation (fig. 2.5).

We can, however, fully grasp Boltzmann's principle only when we also look at processes of motion, because motion, too, is related to the number of practicable possibilities. A teacher's desk often is in seemingly great disarray. But when the cleaning lady has "straightened up" the "mess," the teacher will be most annoyed when he arrives the next day because—as he claims—he can no longer find what he is looking

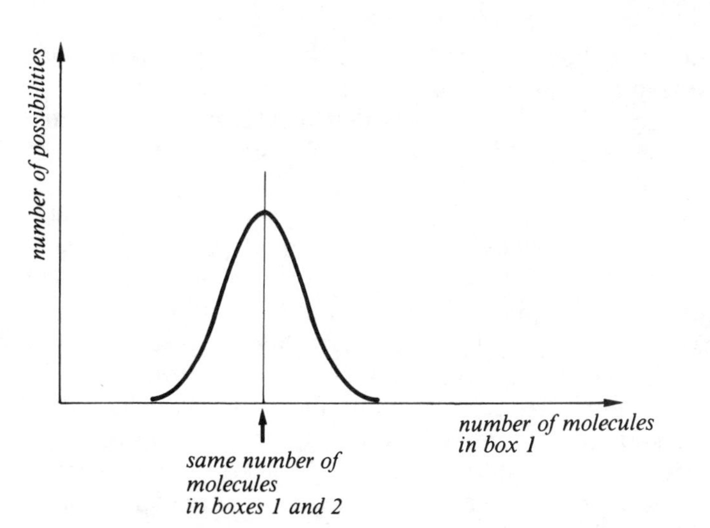

FIGURE 2.5 A so-called distribution curve of a very large number of molecules. The peak of the curve indicates the state when the gas molecules are equally distributed between two vessels. If the distribution is unequal, the number of different possibilities decreases quite rapidly.

for, which he could before the straightening up. Why? Is this merely a whim of his or does he really have a reason for his complaint?

Here is the explanation of this contradiction: On the desk, seemingly untidy to the "uninitiated," the professor knew exactly where to find a certain book or a manuscript page. In spite of the apparent disorder, here too only *one* certain state exists in which he can find his papers. But when the cleaner straightens up, she creates a new state, and the teacher no longer finds the objects in their "right" place. The concept of disorder, then, also includes the continuously new realization of the various possibilities that were demonstrated when we discussed the distribution of the molecules in the boxes; in other words, the situation when the teacher's desk will be in disorder whenever the objects on it are being disturbed.

Nature treats its gas molecules no differently. Because they (for instance, oxygen at room temperature) dart about at 460m/sec, they are continuously stirred up—ever new distributions of the molecules over the two boxes are being realized. Nature is like a poker player who shuffles his cards in front of us at great speed, without our being able to follow his action in detail. The motion with which the molecules occupy always new places is itself irregular—it is the thermal motion.

Energy—Progressively Devalued

These realizations can also be differently formulated, with the example, say, of the automobile. When the automobile moves, its entire energy consists of locomotion; i.e., it is kinetic. Because the locomotion occurs in a certain direction the automobile has, as the physicist says, a single degree of freedom. When the brakes are being applied, kinetic energy is converted into heat, and the brakes and tires become hot (fig. 2.6). Heat, however, means micromotion of very many atoms or molecules. As we all know, one body is hotter than another because its individual molecules move more violently than those of the colder one. But because the molecules can move in various directions at least in the microrange, and are in addition very numerous, heat energy is now distributed among very many degrees of freedom. In other words, when the brakes of the automobile are applied, the energy of a single degree of freedom is distributed among that of very many degrees of freedom, again with a great many possibilities of arranging these distributions. The reverse event would mean that suddenly, as at a command, all molecules would fly in the same direction, and the many degrees of freedom would be reduced to a single one. But this is impos-

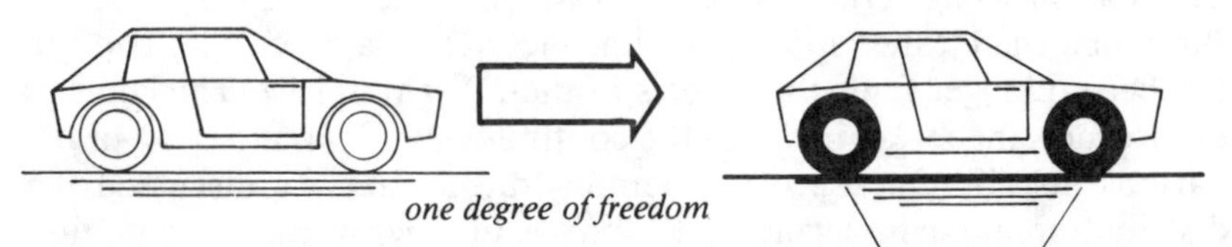

FIGURE 2.6 A moving automobile (*left*) has only one degree of freedom. During braking, this one degree of freedom is converted into the unusually many degrees of freedom of thermal motion, for instance of the wheels and brakes.

sible, according to the basic law of thermodynamics. Although we can convert the energy in the one degree of freedom, the kinetic energy of the automobile, into heat, we cannot reverse this process, at least not completely. As we shall soon discover, the energy concentrated in a single degree of freedom has greater value than when it is distributed among many degrees.

Certain limits can be imposed on nature's tendency toward ever-increasing disorder. For instance, by inserting a partition in a vessel we can prevent a further distribution of the molecules. We can always thus bear in mind that nature need not inevitably achieve maximum disorder, and limitations can be imposed on it from outside. With the aid of tricks, man has succeeded with technology in converting part of thermal energy into useful energy. He has achieved this in the internal combustion engine, for example, by making the piston movable. The heat motion generated by the explosive combustion of gasoline is partly converted into the single degree of freedom of the piston movement, but the major part of the thermal energy is lost and dissipated in the water of the radiator. Physics shows that there are fundamental limits to the "recovery" of high-value energy—moreover, that this calls for machines invented by the human brain and built by human hands. No such limits appear to exist in the universe to prevent the spread of disorder. This has led the physicists to the conclusion that the world is heading toward a state of maximum disorder, in which ultimately all states of order will disintegrate and life will no longer be possible; the world will die a "thermal death." As the famous Hermann L. F. von Helmholtz (1821–1894) expressed it: "From then on the universe is condemned to a state of eternal rest." The equally famous Rudolf J. E. Clausius (1822–1888) said: "The more closely the universe approaches the limiting state in which there is maximum entropy, the fewer will be the chances of further changes." When this stage has been reached, "the universe will be in a state of unchangeable death."

Looking into the past of the universe no more gives us an indication of the chances of life than looking into the future. Almost every physicist believes that the universe originated as an unimaginably hot ball of fire in which there was no order, in a "Big Bang" ten billion years ago. Chaos, then, not order, not only at the end but also at the beginning of the world. And thereafter chaos is assumed to increase even more, up to its maximum. Is there any room left for orderly, meaningful structures, for life as such?

Chapter 3

Crystals—Orderly but Inanimate Structures

In the last section we pointed out that higher temperatures result in a more violent thermal motion of molecules and therefore in greater disorder. This suggests that it might be possible to create an orderly state by removing thermal energy from a system. Indeed, this happens during cooling, as everyday experience shows. When we cool water sufficiently it will turn into ice; more accurately, an ice crystal will form (fig. 3.1).

Because the individual water molecules are minute, only about one-millionth of a millimeter thick, we cannot see them even in the most powerful microscope. But with X rays or electron waves, crystals can be scanned accurately enough to enable the physicists to draw a highly detailed picture of their structure. In a crystal the individual molecules are arranged "in rank and file"—it is a very orderly and at the same time rigid state of matter. In water the molecules are able to slide past each other, which makes the water flow. If we heat the water to its boiling point it will evaporate; in the water vapor the molecules dart wildly about like many tiny tennis balls, colliding with each other all the time and changing direction as they do—they represent a state of total disorder (fig. 3.2).

In physics these different aggregate states—solid, liquid, gaseous—are also called phases, and the transitions between them are called phase transitions. Because these phase transitions obviously create

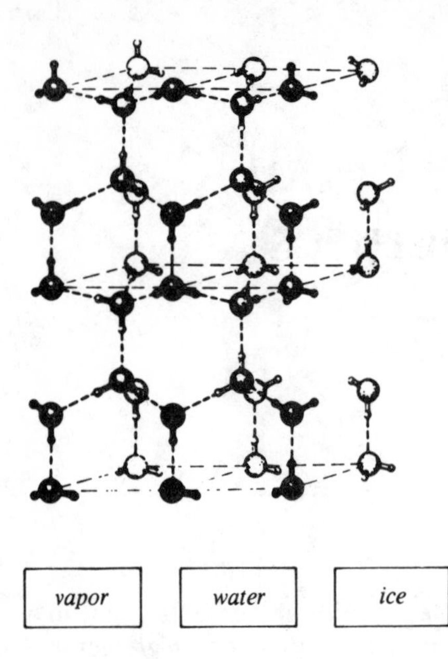

FIGURE 3.1 In the ice crystal the water molecules are periodically arranged in a rigid lattice. The large ball represents the oxygen atoms; the protruding arms, the hydrogen atoms.

FIGURE 3.2 The various states of aggregation of water.

states of widely differing orders or disorders, they have long fascinated physicists. They are still being investigated. What are their special features?

As the example of water demonstrates, the various phases—water vapor, water, and ice crystal—contain exactly the same molecules. Microscopically the three phases differ only in the arrangement of the molecules. In water vapor they move violently at random and at high velocity (about 620m/sec). Practically no forces act between them except during collisions. In the liquid phase the atoms approach each other very closely and are subject to forces of attraction. But the molecules can still move relative to each other. In the crystal, however, the individual molecules are arranged in a strictly periodic "lattice" (fig. 3.3).

Radically different macroproperties are associated with these different microstates of order. The differences in the mechanical properties are particularly striking. For instance, a gas (or water vapor) can be readily compressed, unlike water, which is almost incompressible, and

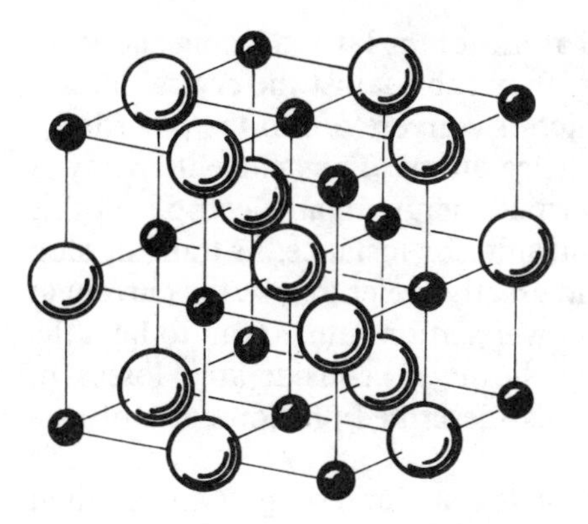

FIGURE 3.3 Arrangement of the atoms in a common salt (NaCl) crystal. Large balls, chlorine ions; small balls, sodium ions.

ice is a solid body. Other physical macroproperties, such as light transmission, also change. These examples illustrate how microchanges can produce completely new macroqualities of substances, not only of water.

Another property of these phase transitions deserves to be mentioned. In given conditions, such as pressure, they take place at a precisely defined temperature, called critical temperature: water boils at 100°C and freezes at 0° (the centigrade scale is based on the boiling and the freezing point of water). Other substances melt at quite different temperatures—iron at 2081°C and gold at 1611°C, to mention two examples—and evaporate at correspondingly higher temperatures.

Superconductivity and Magnetism. Order in the Microworld Produces Strength in the Macroworld

Such phase transitions need not occur only between various aggregate states. A crystal, too, can have further properties that can change abruptly. Superconductivity is a phenomenon of special interest for uses in technology. To understand its extraordinariness, let us think of the transmission of electric current in electric conductors, from transmission lines to radio. When we pass an electric current through metals, it is transported by the smallest charged particles, the electrons. Most metals form a crystal lattice, in which the electrons move like a gas but collide with the individual atoms of the lattice, losing energy in doing so (fig. 3.4). In other words, they rub against the crystal lattice, giving up part of their energy, which is converted into the disorderly, irregular thermal motion of the lattice atoms. Thus current energy is continually lost in the form of thermal energy. In an electric iron this happens to be a desirable effect but not, for instance, in transmission lines, from which consumers would greatly prefer to take the current at the full strength generated in the power station rather than to heat the transmission lines. Unfortunately, however, considerable losses of energy are sustained owing to the just-described phenomenon of friction, called resistance.

As long ago as 1911 the Dutch physicist Kamerlingh Onnes found that certain metals, such as mercury, lose their resistance completely when cooled below a certain very low temperature (fig. 3.5). He called this phenomenon superconductivity. What is really baffling in this phenomenon is the fact that something entirely new happens. It is not that the resistance merely becomes very low; it obviously disappears completely. This has been proved by experiments in which a piece of wire was formed into a loop, which made it endless. A current in this loop ran for over a year without any apparent fatigue; it was the phy-

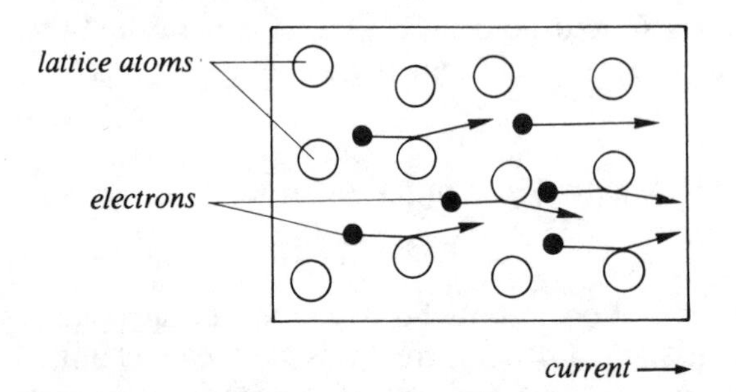

FIGURE 3.4 Microaspect of a metal lattice. The individual atoms of the crystal are represented by large circles. Because of the thermal movements, the atoms continually vibrate. The electrons (small black dots) collide with these vibrating atoms, are ejected from their orbits, and braked. They yield part of their energy to the metal atoms, which heats the metal. At the same time, the flux of the electrons is reduced.

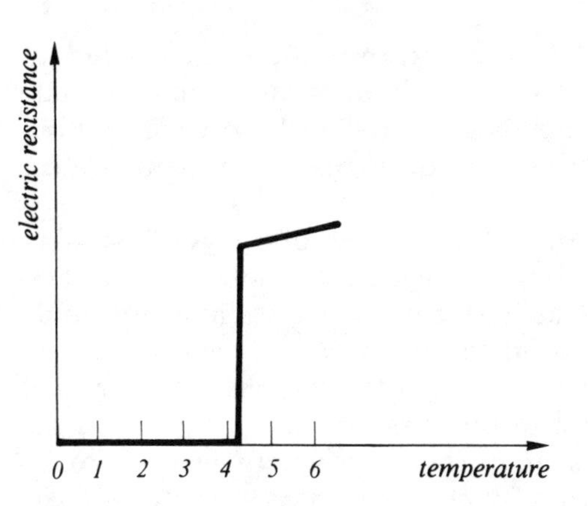

FIGURE 3.5 The temperature is entered horizontally; the electric resistance, vertically. Below a critical temperature (here 4.2° Kelvin [absolute temperature]), the electric resistance disappears completely; above it, the resistance has a finite value.

sicists who became bored, and they allowed the wire to warm up. We had to wait more than forty years for the theoretical explanation of the phenomenon. We now know that the process of superconductivity too is based on a very special microstate of order, in which the electrons of a metal pass through the crystal always in pairs; these in turn combine in a strictly regulated motion that can overcome the attempts of the crystal atoms to offer resistance. We can compare this situation to that of a marching column, arms linked, passing through undergrowth. The undergrowth is unable to deflect the individual marchers from their path. Here, too, as in other phase transitions, a new micro-order ("running in pairs") is associated with an entirely new macrostate (the unresisted current).

Why do we not use superconductors in transmission lines today? The difficulty is that superconductivity occurs only at very low temperatures (for example −260°C), and cooling the cables would be very expensive. But there are other, very important applications where cooling is very much worthwhile. As we know, electric currents can generate magnetic fields. Superconductivity has made it possible to set up enormously strong magnetic fields that are used, for example, in machines that are to produce solar energy on earth by nuclear fusion. Switch elements for computers can be built with the most minute superconductors; the next computer generation will consist of electronic brains that can work only in very low-temperature cabinets, near absolute zero.

Another example of the drastic, abrupt change of a physical property is the iron magnet. Crystals of iron, which are magnetic at room temperature, lose their magnetism abruptly at a certain temperature, 774°C (fig. 3.6). Here, too, it is interesting to trace the explanation on the microlevel. When the physicists dissected the magnet more and more they found it to consist of progressively smaller magnets, the smallest of which are the iron atoms (more accurately their electrons). The "elementary" magnets exert forces among each other. But whereas normally like magnetic poles repel each other, the elementary magnets have the opposite property—i.e., like poles attract each other. In other words, and physically more precisely expressed, all the elementary magnets want to align themselves in the same direction (fig. 3.7). The

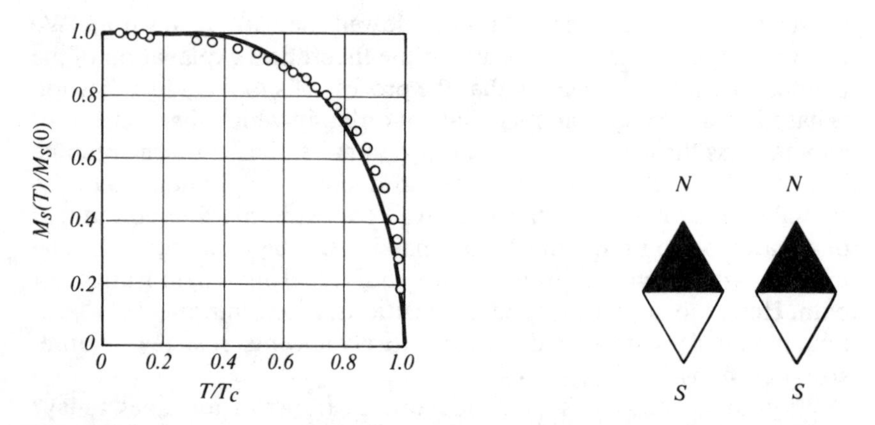

FIGURE 3.6 Again the temperature is entered horizontally; the degree of magnetization of an iron magnet is entered vertically. Above a certain temperature $T = T_c$ the magnetization disappears; i.e., the iron magnet suddenly becomes nonmagnetic.

FIGURE 3.7 The elementary micromagnets of the iron magnet try to arrange themselves parallel to each other, north pole towards north pole, south pole towards south pole.

explanation of this strange behavior became possible only on the basis of the quantum theory, the work of Heisenberg; but to go into detail would divert us unduly from our own subject. All micromagnetic fields add up to generate the macromagnetic field we know from the iron magnet.

Phase Transitions: From Disorder to Order—or the Other Way Around

In the disorderly state of the iron magnet, the elementary magnets can point in all directions. We have, as we say, a symmetrical state. There is no preferred direction. But if the general state is magnetic all elementary magnets point in a certain direction. Although directions are equivalent before the transition, a single closely defined direction is now selected. The original symmetry of the directions is "broken" (fig. 3.8).

The example of the iron magnet shows very clearly how a phase transition takes place in the microrange. In the orderly magnetic phase all elementary magnets are lined up; in the disorderly phase they point in the most varied directions. The cause of these two radically different phases is the competition between two radically different physical forces. One of them aims at the parallel alignment of the tiny magnets, the other is based on thermal motion. Heat, in fact, is disorderly, random motion. Accordingly, thermal motion attempts continuously to push the elementary magnets into the most varied dirctions. We can compare the situation to that of a beam balance. One weight symbolizes the thermal motion, the other the forces of parallel alignment. If the thermal motion "weighs" more it will be the dominant force, the balance will tilt toward one side; i.e., the disorderly motion predominates in the iron magnet (fig. 3.9). The external effects of the individual tiny magnets cancel each other out; we observe no macromagnetization. But if we deprive the iron bar of thermal energy—i.e., if we make this side of the balance lighter and lighter—the forces acting between the magnets will become predominant. The balance will now suddenly tilt toward the other side, and all elementary magnets will align themselves (fig. 3.10).

In the processes we shall presently discuss within the framework of synergetics—not only in physics but also, for example, in sociology

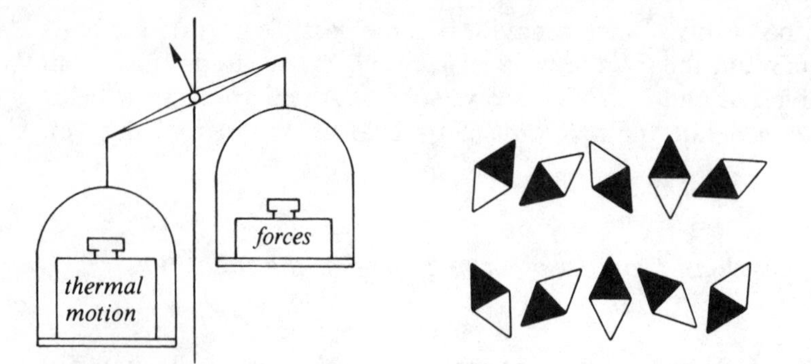

FIGURE 3.8 The balance symbolizes the competition between the thermal motion and the forces that want to arrange the elementary magnets parallel to each other. If the thermal motion is stronger, they will point in all directions.

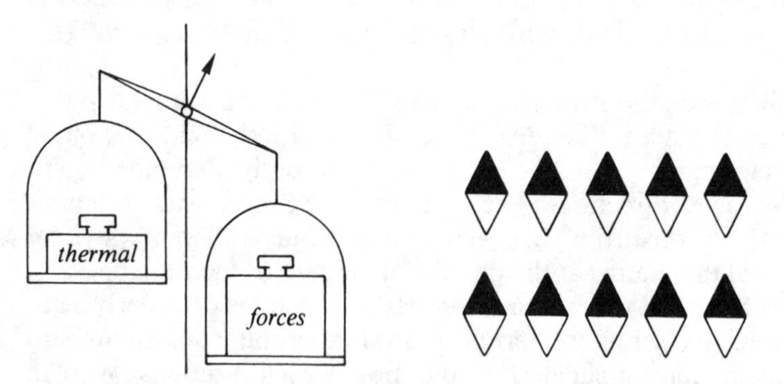

FIGURE 3.9 As in Figure 3.8, but the thermal motion is weaker. The forces dominate and arrange all the elementary magnets parallel to each other.

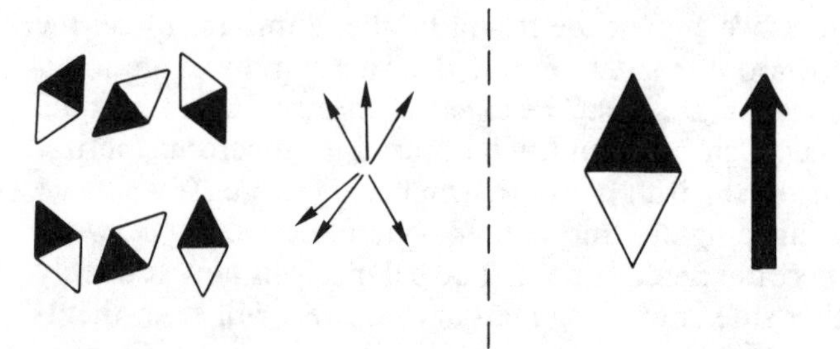

FIGURE 3.10 Figures 3.8 and 3.9 compared. *Left:* If the elementary magnets point in various directions, their external magnetic effect is cancelled; i.e., magnetization is zero. *Right:* If all elementary magnets point in the same direction, their external effect is mutually amplified, resulting in strong magnetization; i.e., the iron magnet is now magnetic.

and psychology—we shall find some concepts important that we may have already encountered in phase transitions.

We can observe such an important property of many phase transitions with the naked eye when a liquid boils. Water below the boiling point, for example, is transparent, as it also is in its vapor phase. When we heat the water slowly it turns turbid, milky, as it approaches the boiling point, because the light passing through it is strongly scattered by the motion of the molecules, which becomes particularly violent near the transition point. Critical fluctuations, as the physicists call them, occur. We can liken the situation to the end of a meeting in a public place. The people suddenly disperse; a violent movement is the result. The crowd becomes denser here, thins out there, and in the end every individual goes his or her own way (fig. 3.11).

As we have already pointed out, the phase transitions are still objects of concentrated physical research. Surprisingly it has been found within the last few years that, despite the different characters of the substances and of the phenomena, these phase transitions obey the same laws; the same basic events, such as the critical fluctuations and the symmetry break, occur time and again. Also within the last few years, physicists have succeeded in establishing the uniform laws.

The sudden occurrence of orderly structure in the phase transitions could obviously tempt us to apply these phenomena directly to life processes because here, too, we are in certain respects faced with orderly systems. There are, however, important objections to this. In the examples we considered, for instance, the substances assumed their orderly state only with a lowering of the temperature. Life processes, however, slow down as the temperature drops; in fact they stop com-

FIGURE 3.11 When a meeting disperses, strong fluctuations in the distribution density of the participants occur.

pletely at very low temperatures and in many creatures even end in death. Living organisms are kept alive by a constant supply of energy and matter, which they take up and process. The more highly developed creatures—i.e., the warm-blooded animals—are not even in thermal equilibrium with their surroundings; quite the contrary. Our body temperature is 37°C; normal room temperature is about 20°C. Life processes obviously must be based on completely different principles, which have nothing to do with the reactions of superconductivity in the crystal formation or of ferromagnetism. Physics can apparently make no contribution at all to the explanation of life. But let us not make a premature judgment; instead let us study the following chapters.

Chapter 4

Patterns of Liquids, Aspects of Clouds, and Geological Formations

We all know that there are several types of equilibrium in mechanics (figs. 4.1–4.3). If we imagine a ball in an open bowl, the ball will be at rest at the lowest point; it will be in equilibrium. If we dislodge the ball,

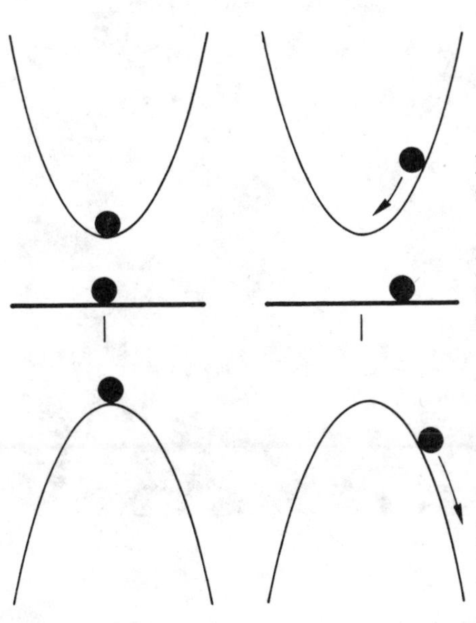

FIGURE 4.1 The ball in a bowl is in stable equilibrium.

FIGURE 4.2 The ball on a level support is in indifferent equilibrium.

FIGURE 4.3 The ball on an inverted bowl is in unstable equalibrium.

it will return to equilibrium. Here its equilibrium is stable. Let us place the ball on a level table; when we move the ball, it will remain at rest in its new position. Here its equilibrium is called indifferent. If we succeed in balancing the ball on the top of an inverted bowl, the ball will again be in equilibrium; but if we push it a little off the top, it will move more and more away from it; here we have unstable equilibrium. We shall need these simple concepts to lead us to a better understanding of certain interesting phenomena of the motion of liquids, phenomena we are all familiar with but seldom aware of. For example, occasionally we may observe in the sky "streets" of clouds, as the experts call them, in strictly arranged formations (fig. 4.4). Glider pilots know that these formations are not static but moving masses of air, with the air rising along one street, sinking along the other. The air thus moves in rolls. Such motions can be produced on a much smaller scale in the laboratory, using a liquid instead of air.

If we heat a layer of liquid in a dish from below, this is what will happen (fig. 4.5): if the temperature difference between top and bottom

FIGURE 4.4 "Streets" of clouds.

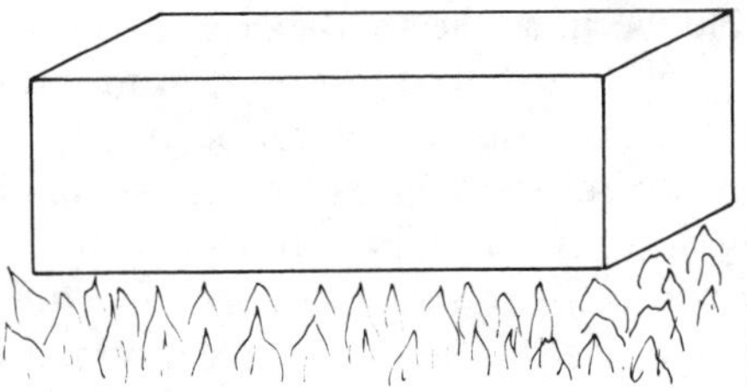

FIGURE 4.5 A layer of liquid heated from below.

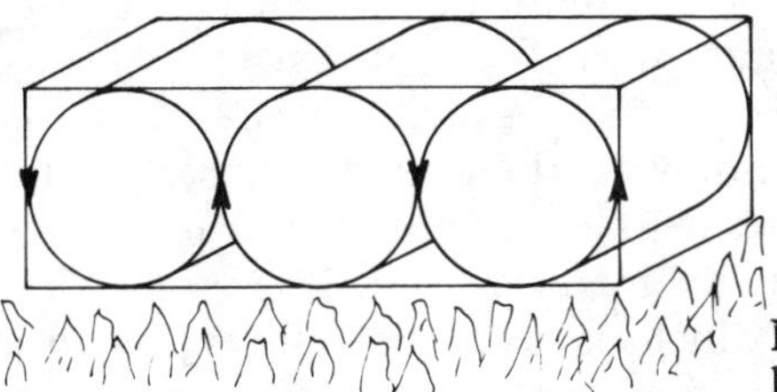

FIGURE 4.6 Rolling motion of the liquid.

is only slight, there will be no macromotion of the liquid. Naturally the liquid will try to equalize the temperature differences by the convection of heat, but as we know heat is a micromotion that we cannot see.

When the temperature difference is further increased something totally surprising will happen. The liquid begins to move macroscopically, not at all in a random fashion but in a quite orderly manner in the form of rolls (fig. 4.6). It rises in longitudinal bands, cools on the surface, and sinks to the bottom. The surprising feature of this roll formation is that the molecules of the liquid must, as it were, communicate across what are huge distances to them in order to establish a collective motion. The liquid rolls are many billions of times larger than the individual molecules of the liquid. To begin with, let us look at a liquid layer at rest. If it is heated from below, its lower portions naturally expand and want to rise. The colder and therefore heavier portion of the liquid presses down from above. But the portion of the liquid that wants to rise is in equilibrium with that which wants to sink (fig. 4.7). Is this equilibrium stable or unstable? At first glance it would appear that the position is unstable, because the top layer wants to go down, the bottom one up; all we would have to do to set it in motion would be to give the liquid a slight push. But the situation is a little more complicated, as we shall soon discover. Let us imagine a small heated sphere of liquid rising (fig. 4.8). As it makes contact with the colder layers, heat will be conducted away from it. It will become cooler, contract, and lose the tendency to rise. In addition, its motion is retarded by friction from the surrounding matter as it rises. Owing to cooling on the one hand and slowing down on the other, the liquid sphere ceases to rise; the liquid layer must therefore remain at rest after all. This situation, however, can be maintained only if the temperature differences are not too great, because if the liquid is sufficiently heated, the hot drop of liquid will succeed in rising and induce macromotion. The astonishing fact is that such hot drops do not rise irregularly but in an even, orderly manner. It looks as if an external power is at work, which we can easily recognize by an analogy.

Imagine a swimming pool in which the swimmers are supposed to swim in one direction to the other end and back. If the pool is full, which will happen on hot summer days, very many swimmers will be in it and get in each other's way—obstruct each other—as they swim back and forth (fig. 4.9). Some life guard will therefore suggest that the swimmers should move in a circle (fig. 4.10). Obstruction will thus be

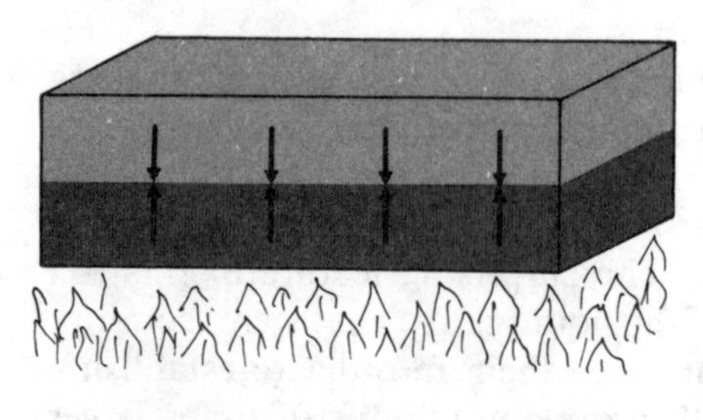

FIGURE 4.7 Liquid still at rest.

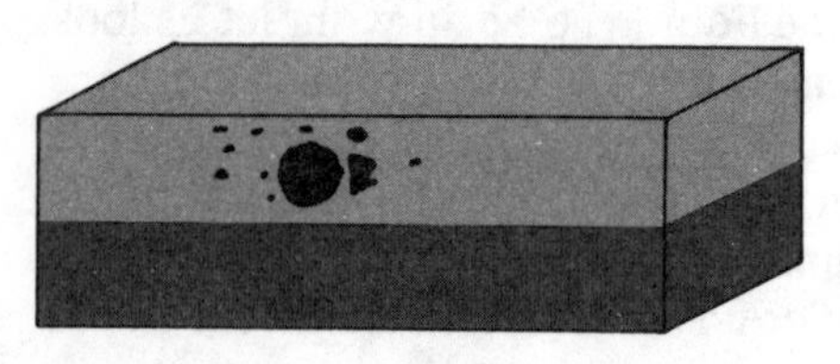

FIGURE 4.8 Rising sphere of liquid.

FIGURE 4.9 Swimmers in a pool. Uncontrolled movement.

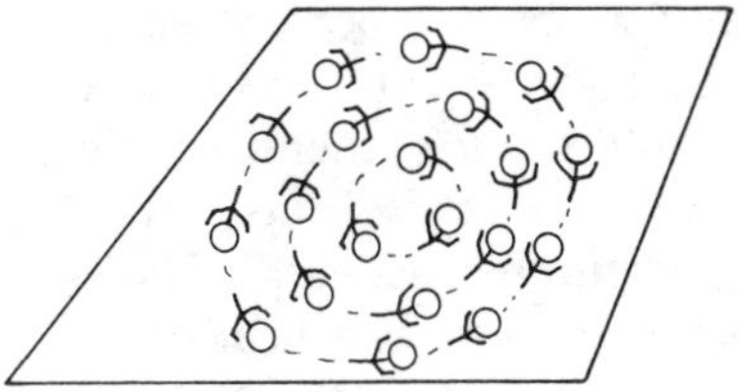

FIGURE 4.10 Swimmers in a pool. Circular, controlled movement.

greatly reduced. Here the life guard has imposed on the swimmers a collective motion. But even without benefit of the life guard, the swimmers may get the idea of moving in a circle. Only a few of them to begin with, perhaps, but more and more will follow as they find swimming in a circle more convenient. Eventually a collective motion comes about without any external instructions; i.e., it is *self-organized.*

Nature, here the liquid, acts in the same way. It "discovers" that it can transport the heated parts upward much more efficiently when they join in a regular motion. But how does the liquid manage it? Through fluctuations. The liquid continually tests different possibilities of motion by time and again dispatching small hot quantities of liquid, on trial as it were, and letting cooler ones sink down in return. In our imagination we can see these most varied possibilities of motion split up into particularly simple motions; any violent motion of a liquid can be dissected into seemingly uniform types of motion; two of them are illustrated in figures 4.11 and 4.12. In one the liquid finds that the conditions are particularly favorable for the rise of the hot portions. This type of motion continuously increases; more and more parts of the liquid are sucked into this motion, are "enslaved" by it. After some time the other type of motion subsides. It was only a kind of fluctuation. Here we encounter the competitive behavior of different collective types of motion: one of them becomes more and more dominant, suppressing as it does so all the others. This rolling motion plays the part of an order parameter. It directs the motion of the individual parts of the liquid. Once such a form of motion is established even in only parts of a liquid, other parts of it are sucked into it, or, in other words, they become enslaved by the order parameter. It is interesting to note that we can precisely calculate which collective motion will eventually gain the upper hand, and which will become enslaved by it. However, this must be taken with a grain of salt. If we look at an individual roll, say the middle one, it will be obvious that basically the motion can either be clockwise or counterclockwise (figs. 4.13–14). Which of the two directions is selected is purely accidental. The symmetry of the clockwise and counterclockwise motions is broken by the accidential initial fluctuation. Once the original state of rest of the liquid's motion has become unstable, a very minor fluctuation will be enough both to start the rolling motion and to determine the macromotions. We shall in due course find in sociology that in political or economic decisions, minor fluctuations—accidents, as it were—often determine the momentous direction events will take. Once the choice is made the alternative is out of the question, and the choice cannot be reversed. Minor fluctuations often decide the nature of the choice. Once it has been made, all particles must willy-nilly accept it.

At the beginning of this section we explained the various types of equilibrium with the aid of a simple mechanical model, a ball and a bowl. The stabilization of the rolls can also be understood with the help of such an idea. Let us enter the maximum vertical velocity to the right; we thereby demonstrate the magnitude of the velocity by the displacement of a ball. When the rest position of the liquid is stable all fluctuations of this velocity must subside to zero, which is illustrated in figure 4.15. If we increasingly heat the liquid from below, the state of rest will be unstable. With a small fluctuation the vertical velocity will increase. We can visualize the new unstable situation, as at the beginning of this section, in figure 4.16. But because the rolls eventually become stable the velocity is no longer permitted to increase; it has reached a stable, final value. The ball has returned to the bottom of the bowl. Figure 4.17 represents a combination of the pictures. But because the clockwise and counterclockwise directions of the motion are equivalent, the aspect of our figure must be symmetrical; i.e., figure 4.18 applies to the velocity, symbolized by the position of the particle.

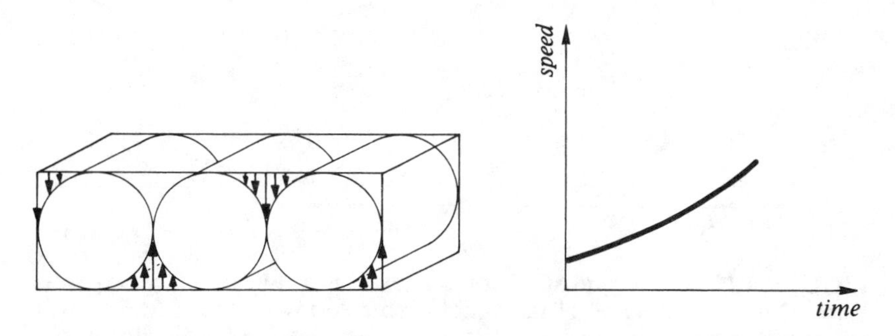

FIGURE 4.11 *Left:* Possible arrangement of the movement rolls. *Right:* In time, the rolling speed (vertical) progressively increases.

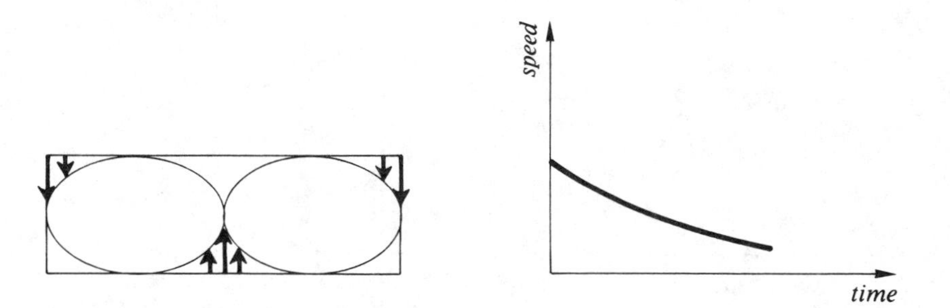

FIGURE 4.12 Another configuration of rolls, whose speed of rotation slows down in time.

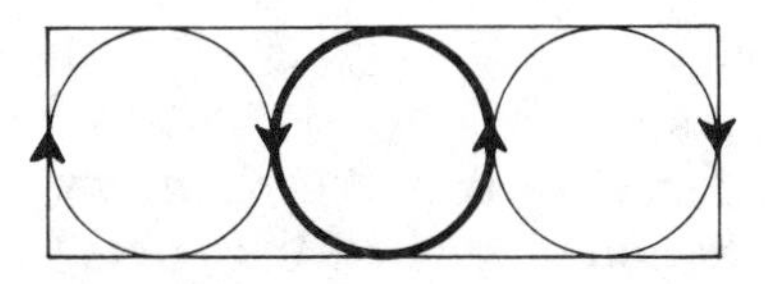

FIGURE 4.13 Breaking of symmetry. Here the center roll rotates counterclockwise.

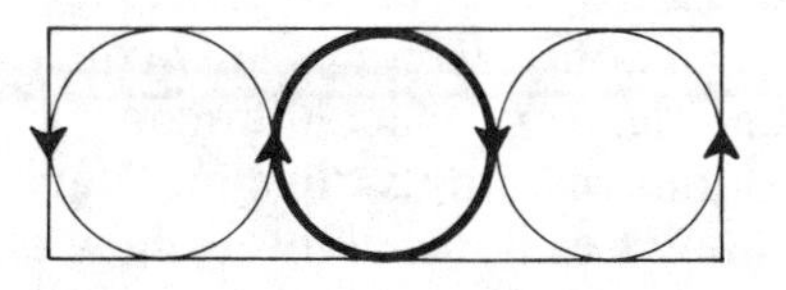

FIGURE 4.14 As in Figure 4.13, but the center roll rotates clockwise. Correspondingly, the rotation direction of the other rolls is reversed.

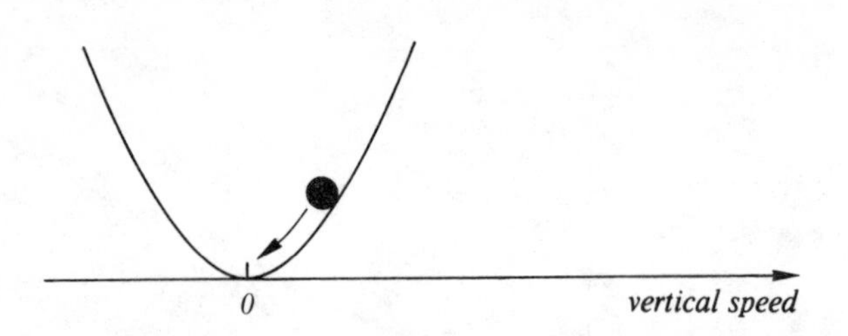

FIGURE 4.15 Equilibrium state of the liquid when it is heated weakly from below. The vertical speed of the liquid is entered toward the right. The ball, whose position symbolizes the vertical speed, returns to its rest position during a disturbance.

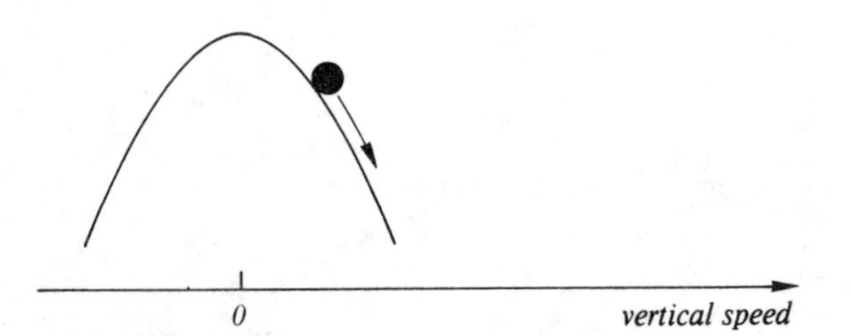

FIGURE 4.16 If the temperature difference between the bottom and top boundaries of the liquid is sufficient, the speed of the rolls will increase. According to our mechanical analogy, the position of the ball is unstable.

We can thus clearly see the already described break in the symmetry, but in a different way. The ball, whose position symbolizes the velocity of the rolls, can assume basically two equivalent positions but must obviously decide in favor of one of them and thus break the symmetry. Rolling motions are not the only possible macromotions of liquids heated from below. If the liquid is contained in a round vessel, the axial direction of the roll is completely random. Here, not only competition between the various rolls is possible, from which ultimately a single direction will emerge; rolling motions in various directions can mutually stabilize each other. The best known example of this is shown in figure 4.19, where the rolling motions mutually support each other

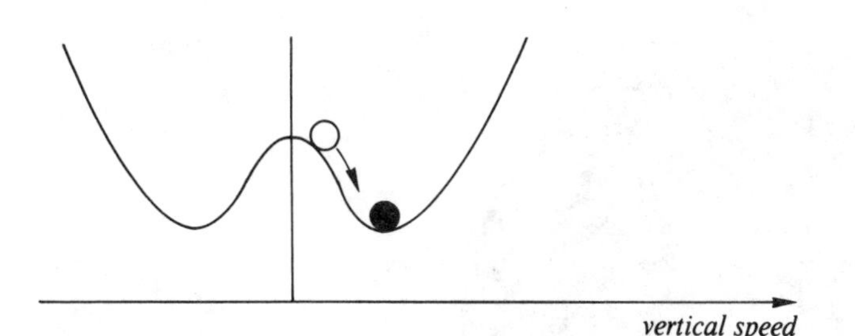

FIGURE 4.17 Because the vertical speed does not increase infinitely, but eventually becomes zero, the unstable position of the ball in Figure 4.16 must finally become stable as shown here.

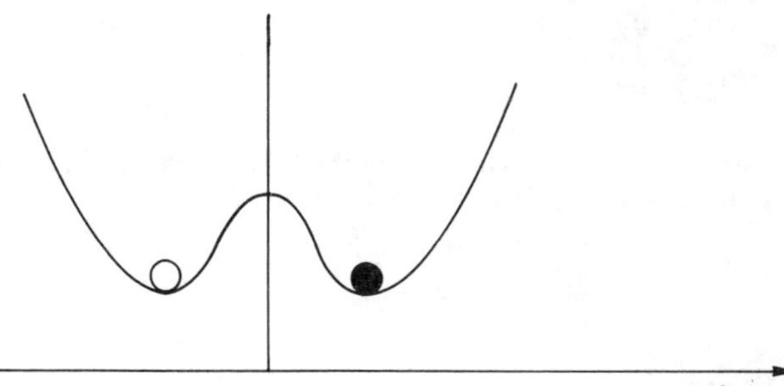

FIGURE 4.18 Breaking of symmetry. The ball can assume one of two equivalent positions, and the rolls can rotate either clockwise or counterclockwise.

and stabilize themselves like three mutually supporting poles which, as a result, remain upright (fig. 4.20). If we add the individual motions of the rolls, which is somewhat laborious, we obtain a honeycomb—i.e., a hexagonal pattern. The liquid rises in the center of such hexagons, and sinks along the outside. When, for example, a round can of ski wax is heated from below, this is the pattern that will be produced.

The example shows immediately that the concept of liquid is wide-ranging. Even lava forms hexagonal blocks as it solidifies. In salt lakes heated by the earth's interior, more or less regularly hexagonal sheets of salt may crystalize out; if red bacterial cultures grow on them they produce the formations shown in figure 4.21.

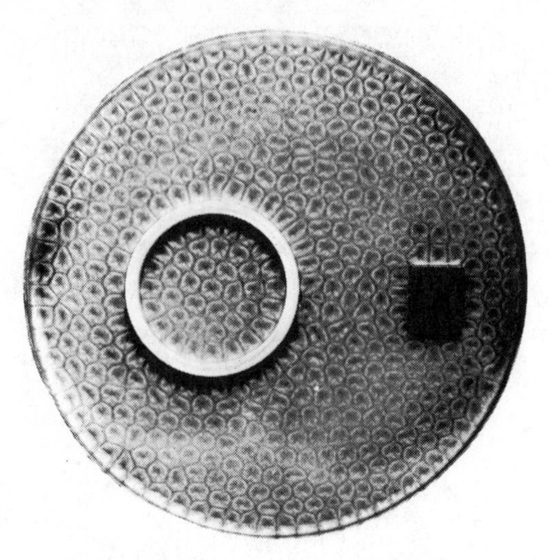

FIGURE 4.19 Honeycomb pattern of the motion in the liquid. The liquid rises in the center of each honeycomb and falls along the walls.

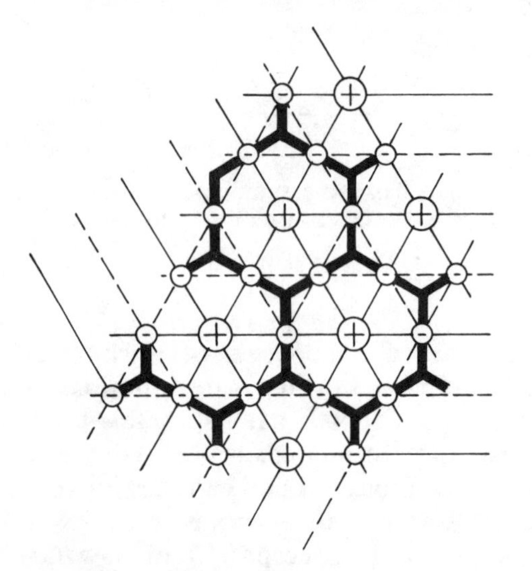

FIGURE 4.20 Creation of the hexagonal honeycomb pattern of Figure 4.19 owing to the superimposition of rolls of different orientations. In the circles marked "+" the liquid rises; in those marked "−" it falls. The full and dashed lines represent the boundaries of the rolls. Along the full lines the liquid rises; along the dashed lines it falls. The bold lines indicate the incipient hexagons where the liquid moves downward.

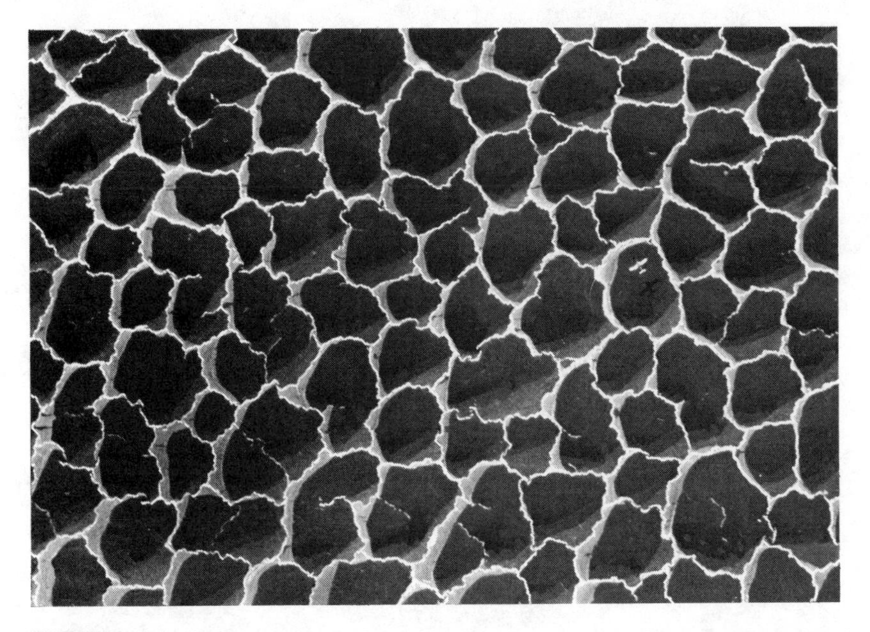

FIGURE 4.21 Hexagonal salt deposits. Purple bacteria color the dry bed of Lake Natron in East Africa.

Astronomers observe structures on the surface of the sun which are called granules. It is assumed that they are the result of this phenomenon (fig. 4.22). If a liquid in which a honeycomb pattern has formed is heated more strongly from below, this pattern can be displaced by simple rolls; i.e., figure 4.6 instead of figure 4.19 will illustrate the situation. The mathematical analysis, into which we can obviously not enter here, admits of an interpretation which is both amusing and thought-provoking; in the new conditions competition begins between the original three directions of the rolls, which had mutually stabilized each other to form the honeycomb structure. This competition is won by one of the rolls, again through an accidental fluctuation; it will become dominant and enslave the others—i.e., subject their motion to its own.

As we have seen in this kind of description, the demonstration of natural phenomena here merges into those that play a part in psychology and sociology. But the natural reactions described here have the

FIGURE 4.22 Granules on the sun.

advantage of being amenable to mathematical analysis and examination.

It is utterly surprising that very diverse natural phenomena should be subject to exactly the same laws; many more examples of this will be described later. But what we already know about the subject allows us to establish the principle even at this stage. If we change external conditions, such as the temperature difference between the bottom and the top layer of a liquid, the old—for instance the rest—state will become unstable and will be replaced by a new macrostate. Near the point of transition the system tests new possibilities of an orderly macrostate with continuous fluctuations. At the point of instability itself and a little above, the new collective form of motion will become progressively stronger, finally gaining the upper hand over all the others. With some of these collective motions, competition is not the only reaction. It may come to a cooperation between equivalent forces, which produces new patterns. In contrast with the phase transitions in thermal equilibrium, however, kinetic patterns are created time and again; i.e., we always encounter dynamic situations. The shape of the outlines sometimes aids the formation of the patterns. If the outlines are let us

FIGURE 4.23 Top view of a layer of liquid heated from below in which two rolling motions vertically to each other are forming.

say vertical, two forms of rolls vertical to each other can coexist, producing the pattern of figure 4.23.

Even more complex patterns are observed; see figure 4.24. They are no longer static but visibly in continual motion, and constantly show pulsation of the liquid motion, which sometimes looks almost as if the liquid were breathing.

Scale of Motion Patterns

But motion patterns in liquids are set up not only by heating. The following experiment is relatively easy to perform in the laboratory. Fill an outer cylinder containing a coaxial inner one with a liquid. Rotate the inner cylinder. As it rotates, the inner cylinder naturally takes part of the liquid with it; another part adheres to the wall of the outer cylinder. Concentric flow lines are produced at low speeds of the inner cylinder's revolution; but if the speed is raised above a critical value, a completely different motion, rolls as shown in figure 4.25a, will begin. These rolls are "packed" like bent sausages in a can. When

FIGURE 4.24 Top view of a layer of liquid heated from below in which a complex motion pattern closely resembling a carpet design has developed.

the speed of revolution is further increased the rolls begin to vibrate, waves begin to circulate (fig. 4.25b). At even higher speed these vibrations will become more complex (fig. 4.25c), to be converted into a totally irregular motion at a final stage of speed. This motion is called turbulence, and most recently also chaos (fig. 4.25d).

As this example of liquid motion reveals, increasingly complex motion patterns can be created by self-organization. In the language of synergetics, new order parameters continuously succeed each other.

The generation of an entirely irregular, chaotic motion might lead to the assumption that the order parameters have lost their power to control. We shall give the answer in Chapter 11.

This example is so important also because it shows that in strictly defined experimental conditions, even a chaotic motion can be produced during self-organizing events. Within the last few years the exploration of such chaotic motions has greatly expanded. Mathematical models prove that such phenomena can be inevitable not only in physics but also in entirely different fields such as economics. We shall thus find that we are forced to abandon certain dogmas of economic theory. Those readers who at this point would like to infer that self-organization can lead to chaos but that organization—i.e., control

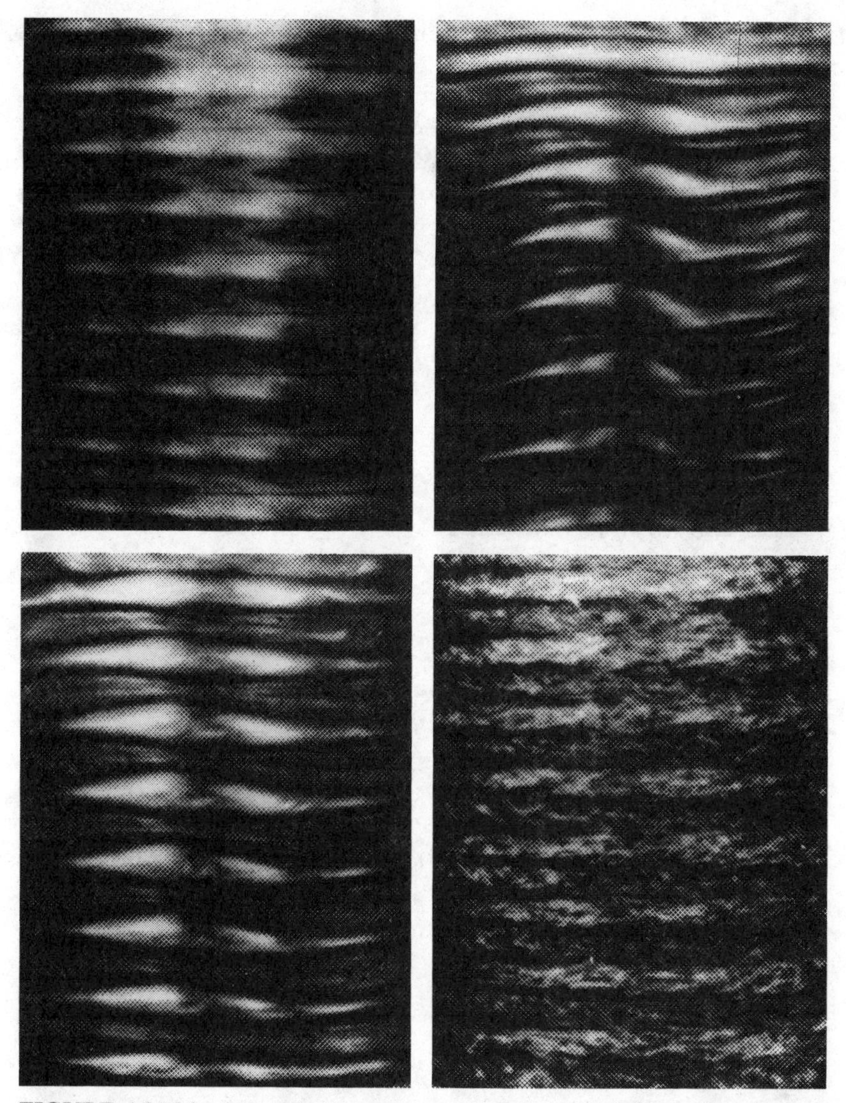

FIGURE 4.25 Liquid motion between two vertical coaxial cylinders. The outer cylinder at rest is transparent; the inner one rotates. Different liquid patterns form according to the different speeds of rotation of the inner cylinder.
a) Rolls surrounding the inner cylinder resembling small sausages.
b) The rolls oscillate.
c) The rolling motion becomes even more complex.
d) Irregular, chaotic motion.

from outside—avoids it should realize that they will find that the very control processes can also produce chaos in self-organizing systems.

But let us briefly return to physics. The occurrence of increasingly complex patterns is a widespread phenomenon in flow dynamics, to be demonstrated in figure 4.26. Here a liquid flows around a cylinder; the speed of the liquid increases from one figure to the next. This creates

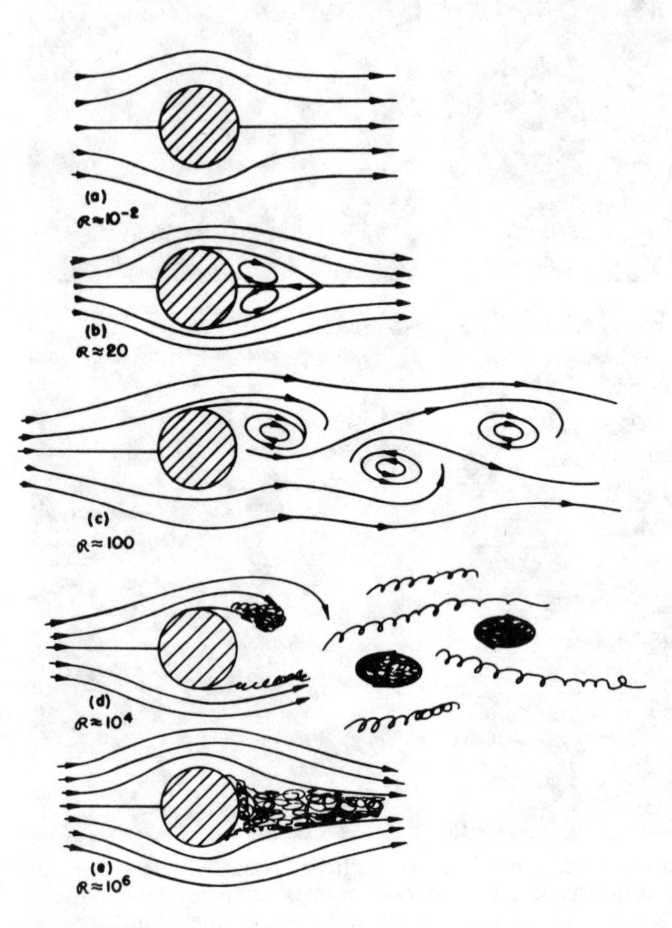

FIGURE 4.26 Streamline aspects of a liquid flowing around a cylinder. As the speed of the flow increases, the flow pattern becomes more and more complex.

different patterns in a strictly defined sequence and all are connected with eddy formation.

All these phenomena will perhaps appear odd and their study trivial. But we have seen at the beginning of this chapter, in the example of cloud formation, that these phenomena also occur on a much larger scale; they even explain the continental drift of the earth's crust. When we look at the globe we see that for example, the east coast of South America fits the west coast of Africa like a glove. Not only this superficial observation but also far-reaching scientific comparisons of geological formations and of the respective flora and fauna led the German geologist Alfred Wegener (1880–1930) to his theory of continental drift. According to this theory, the continents have moved thousands of kilometers on the earth's surface in the course of millions of years. The hypothesis appears very bold indeed, because we are apt to regard the earth's surface as something fixed, rigid. However, we must remember that the interior of the earth is very hot and behaves rather like a viscous liquid. And this immediately gives us our clue. We can regard a layer between the earth's center and its crust as a liquid heated from below, having a certain temperature at the top. This is precisely what sets the convection currents in motion, which move like rolls and are therefore capable of moving whole continents. It is a very slow process indeed.

Similarly, model experiments can be conducted with a rotating glass sphere filled with liquid. Here, too, entirely specific patterns are created in the form, say, of moving bands on the surface of the liquid, which provide models for the various belts of gas on the planet Jupiter.

Theoretical physicists and astrophysicists are able to calculate and predict such pattern formations; the basic phenomenon is always the same: the growth of certain modes, certain forms of motion, which will stabilize themselves on the principle of enslavement.

Chapter 5

Let There Be Light—Laser Light

There Are Different Kinds of Light

In 1960 I acted as a scientific consultant to the Bell Telephone Laboratories in New York City. To a much greater extent than their European counterparts, American companies maintain large research laboratories, in which a very close interrelation between thoroughgoing pure research and effective application is a matter of course. Very soon I was initiated into one of the central secrets of then current research activity, in which several teams were involved. Attempts were being made to develop a source that would produce light of entirely new properties. A paper by Arthur Schawlow and Charles Townes published in 1958 gave the impetus to research. Even earlier, in 1954, Townes together with his coworkers had built a device which generated the so-called microwaves in a completely novel way. Like radio and radar waves, microwaves are electromagnetic. We cannot detect any of these waves with our sense organs; they exist all the same. It is as if we were standing on the seashore in pitch darkness. The waves of the sea will then be invisible to us; we can nevertheless see them when a boat with a lantern bobs up and down on them. The situation is similar with electromagnetic waves, where a radio, for example, produces evidence of their existence. After certain conversions of the electromagnetic vibrations, the radio makes these waves audible to us.

The task at Bell Telephone Laboratories (as well as in the laboratories of their rivals, which in the United States were not exactly asleep) consisted in the generation of light waves according to Townes's principle of microwave generation. This principle had been named "maser," which is an acronym like many other terms of modern science and technology, an artificial word, almost a linguistic game; it is made up of the initials of *m*icrowave *a*mplification by *s*timulated *e*mission of *r*adiation—unintelligible to most of us. The word "laser" was the logical variant: instead of microwave, we wanted *l*ight *a*mplification by *s*timulated *e*mission of *r*adiation.

After this brief excursion into neologisms let us examine the enormous step forward the laser represents compared with the lamp. To appreciate it properly we must first briefly concern ourselves with lamps and with the light they radiate. This will reveal, incidentally, a direct path to some fundamental ideas of synergetics.

Let us take a so-called gas discharge tube as an example of a lamp. It is a glass tube filled with an inert gas such as neon. An individual atom of gas consists of the positively charged nucleus and a number of negatively charged electrons, which orbit this nucleus like planets orbit the sun. To simplify matters we shall investigate only the behavior of one electron, the so-called photoelectron (fig. 5.1). As the Danish physicist Niels Bohr discovered in 1913, an electron can occupy only certain orbits; others are "forbidden" to it. This behavior was explained only by the quantum theory, according to which the electron behaves not only like a particle but also like a wave, which as it were must bite its own tail as it orbits the atomic nucleus. This is why only strictly defined orbits are possible. Normally the electron moves in a valley, in the lowest orbit (fig. 5.2). If we pass an electric current, which is carried by many freely whirling electrons, through the tube, these electrons

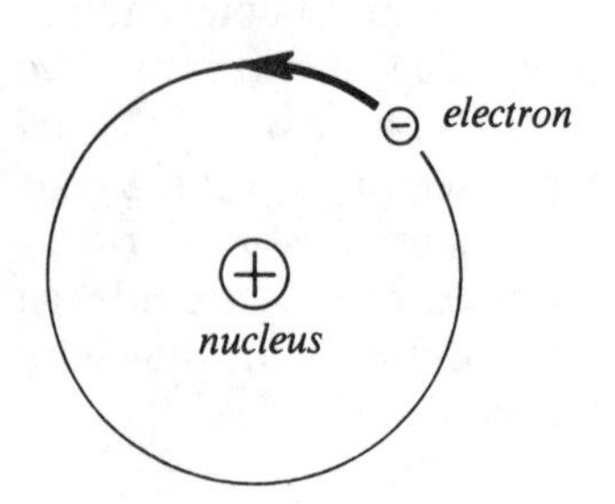

FIGURE 5.1 Scheme of an atom (hydrogen). A negatively charged electron orbits the positively charged nucleus.

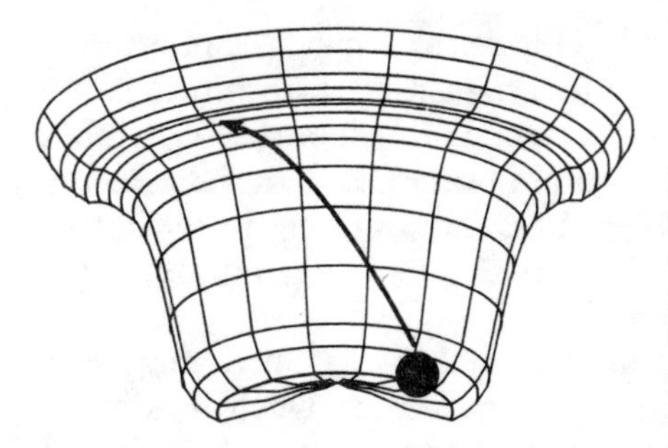

FIGURE 5.2 Motion of the electron (black ball) around the atomic nucleus. The electron moves in groovelike valleys. The supply of external energy—for example, light—can raise the electron from the lowest to a higher groove.

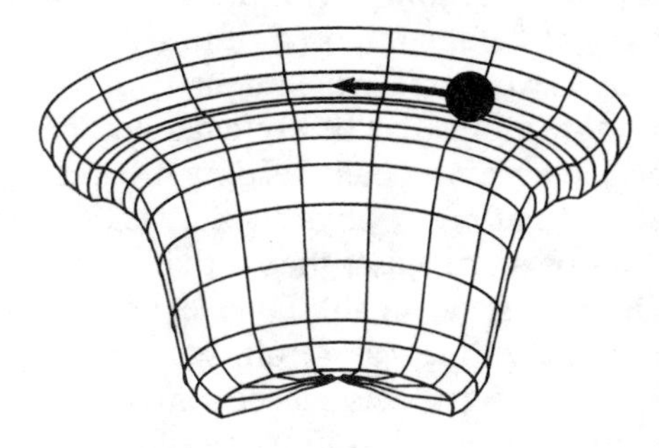

FIGURE 5.3 The electron orbits in the higher groove. This corresponds to the excited state of the atom.

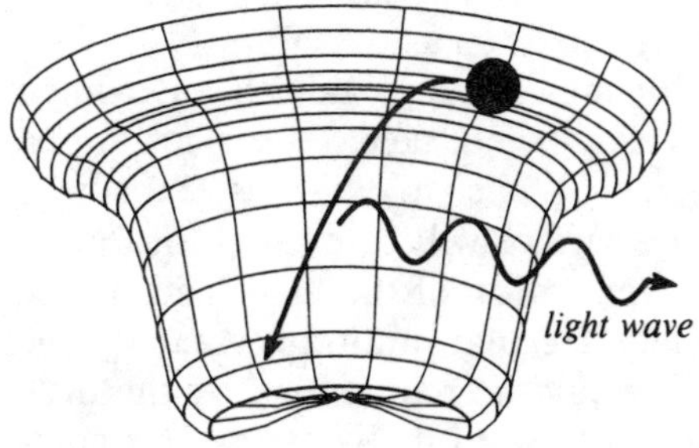

FIGURE 5.4 The electron drops from the higher back into the lower groove, yielding as it does so its energy in the form of a light wave.

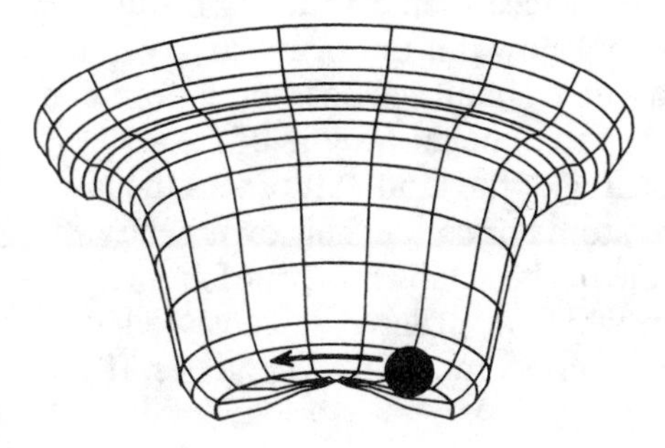

FIGURE 5.5 The electron again orbits in the bottom groove.

will collide with the atoms of the gas. During the impact the photoelectron of an atom can be kicked up into a higher energy orbit (fig. 5.3), from which it can jump back into its original orbit spontaneously, that is quite suddenly, at no foreseeable point in time. It releases the energy thereby liberated to the light field (fig. 5.4), and continues its motion in the lowest orbit (fig. 5.5). A light wave is generated like a water wave when we throw a pebble into a pond.

In a gas discharge tube many photoelectrons obviously meet this fate. Light waves are generated in the same way as when we drop a large number of pebbles into our pond at random. As on the surface of the water, a violent motion of the light field, composed of individual wave trains like spaghetti, will result. If we increase the intensity of the current through the gas, more and more atoms will be excited, and one would expect that the tangle of the wave trains would become progressively worse. This was indeed the view of many physicists. I was the first (I am still proud of the fact) to show in my laser theory that something entirely different happens here. Instead of the tangled knot, a completely uniform, practically infinitely long wave train will result. This prediction was fully confirmed by experiments subsequently conducted in various laboratories all over the world. A drastic difference, therefore, exists between the light of an ordinary lamp and laser light. The following analogy explains why this is so astonishing.

Let us represent the atoms by little men standing along a water-filled canal (fig. 5.6). The water symbolizes the light field. Its surface at rest corresponds to the absence of a light field—i.e., to darkness. If the little men push sticks into the water, the surface will be excited into wave motions. This corresponds to the generation of the light field by the atoms. As with a lamp, a completely irregular motion is created. The laser situation, on the other hand, would be represented by the little men pushing their sticks simultaneously into the water as at a command, so that a regularly moving surface of the water is produced. Within the human sphere it is obvious how this uniform activity of the little men comes about. Behind them stands a boss or a foreman always shouting "now, now, now" so that the pushing-in of the sticks is strictly controlled. But there is nothing of the kind that orders the laser atoms to behave in this way. The atoms organize their own behavior. The laser thus provides an example of the establishment of an orderly state through self-organization, in which disorderly is translated into orderly motion. It makes the laser a perfect model of synergetics. It can

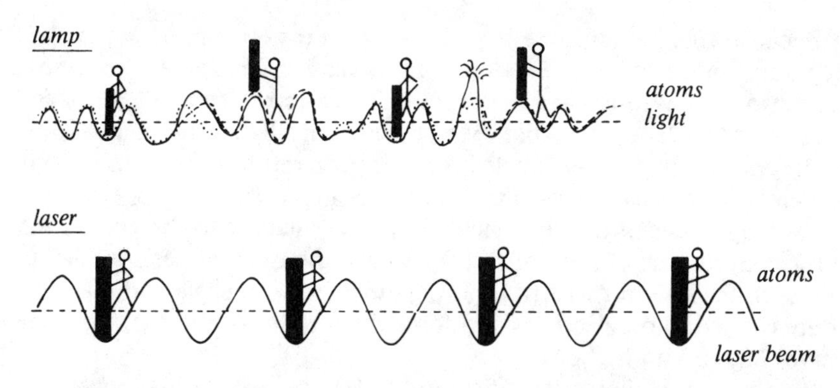

FIGURE 5.6 Function of a lamp and of a laser. The figures holding sticks are standing on the bank of a canal filled with water. *Top:* They are pushing their sticks into the water independently of each other. The turbulent surface of the water corresponds to the light field of a lamp. *Bottom:* The sticks are rhythmically pushed into the water. A regular wave of water, corresponding to the laser beam, is generated.

also serve as an allegory for many processes in entirely different disciplines, especially in sociology.

But before we go further we must delve more deeply into the basic ideas of synergetics, last but not least because it might otherwise seem that we transferred findings in physics directly, in a shallow way, without any further reflection, to phenomena as complex as those of human society. It will nevertheless, be easy to recognize in the laser some basic features that take us a step nearer to the understanding of processes of animate nature.

Self-organization in the Laser

Let us take a closer look at the laser to discover the secret of its self-organization. The laser and the ordinary gas discharge tube differ only in that the laser has two mirrors at either end of the glass tube (fig. 5.7). These ensure that the light, which moves along the axis of the tube, remains in the tube as long as possible (fig. 5.8). If one of the mirrors is made slightly translucent, some of this light can be emitted. But why should we want to keep the light longer in the laser arrangement?

Because this can initiate a process that Einstein predicted at the beginning of the century. Existing light waves can force an excited photoelectron to covibrate at their rhythm. The electron can be likened to a frenzied tap dancer who dances to the beat of a band only to collapse at the end, completely exhausted. The electron reinforces—amplifies is the correct physical term—the light wave; i.e., the electron heightens the wave's peaks until it has yielded its entire energy to the wave and returns to its basic "state of rest." Because the mirrors keep the light waves in the laser for a relatively long time, they can enslave more and more excited photoelectrons and force them to make the peaks of the waves higher and higher.

But even if the height (= the amplitude) of the waves is the same, the waves need not be uniform. In some cases peak follows peak in rapid succession, in others the distance betwen the peaks is greater (fig. 5.9). We indeed find already at the start of every laser transmission entirely different waves in the laser, emitted by a few "rash" photoelectrons. These waves compete with each other in their demand to obtain amplification from the rest of the excited photoelectrons. But the electrons do not amplify various light waves in completely the same way; they yield their energy with usually a very slight preference for a certain wave. It is the wave whose rhythm most closely approaches the inherent "dancing rhythm" of the photoelectrons. Although this particular wave is often preferred only to a very slight extent, it will be amplified, reinforced like an avalanche, and eventually dominate all the others; they will be suppressed, and all the energy of the photoelectrons will be transferred to the one very regularly vibrating wave. Conversely, once this wave has established its dominance, it will always enslave each newly excited photoelectron of an atom and make it covibrate in its cycle. The newly generated wave thus determines the order in the laser; it plays the part of the order parameter, a concept that should by now be familiar to the reader.

Because this order parameter forces the individual electrons to vibrate exactly in phase, thus imprinting their actions on them, we speak of their "enslavement" by the order parameter. Conversely these very electrons generate the light wave—i.e., the order parameter—by their uniform vibration. The existence of the order parameter on the one hand and the coherent behavior of the electrons on the other mutually condition each other. This is a typically synergetic phenomenon. In order for the electrons to vibrate uniformly in phase, there must be an order parameter—the light wave. The light wave, however, is

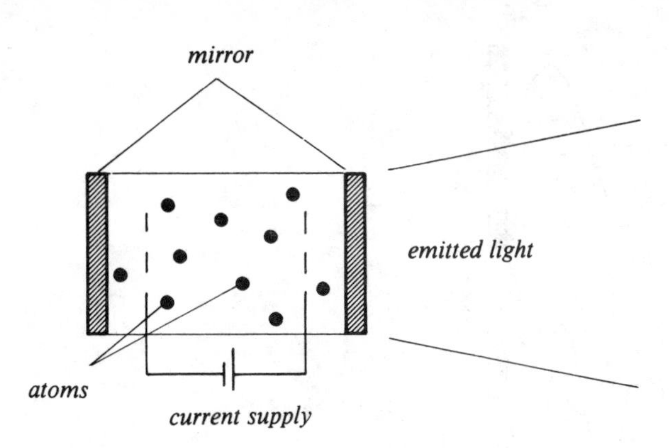

FIGURE 5.7 Typical laser arrangement.

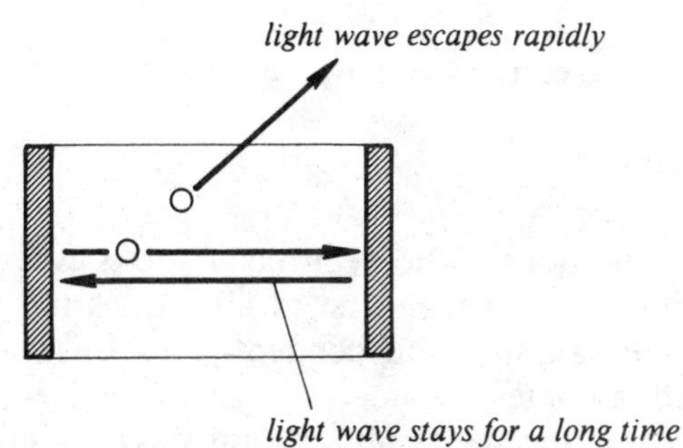

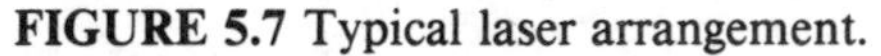

FIGURE 5.8 Different behavior of light waves between two mirrors. The wave traveling obliquely to the axis escapes rapidly; the axial wave remains in the laser for a long time.

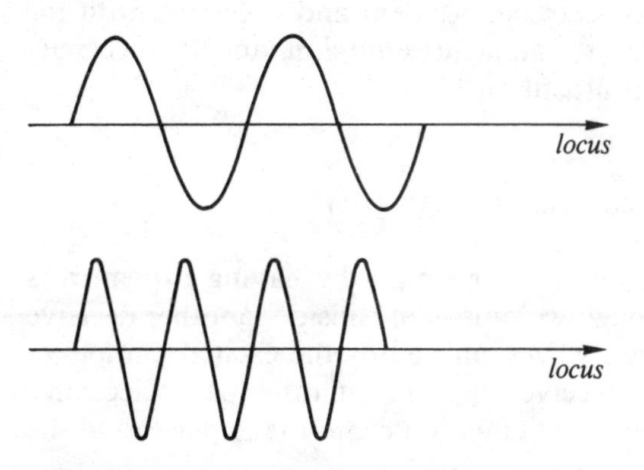

FIGURE 5.9 Waves are not uniform. Two examples in which the distances between wave crests differ.

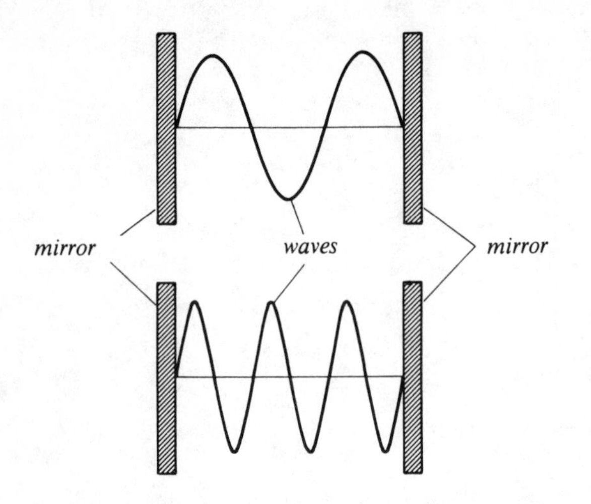

FIGURE 5.10 Only closely defined waves fit between two mirrors.

generated only by the uniform vibrations of the electrons. It looks as if we have to postulate a superior power, which first of all creates the state of order, so that this can maintain itself independently. But this is not so, as we have just learned; a contest, a process of selection preceded it, all electrons have become the slaves of a certain wave. It is interesting to note that initially the individual waves were generated by the electrons entirely by accident, spontaneously, but then were selected, isolated according to the laws of competition. We have the interplay, typical of synergetics, between accident and necessity, with the accident represented by the spontaneous emission, and the necessity by the inexorable law of competition.

The Laser—an Open System with a Phase Transition

But does every lamp become a laser simply by having two mirrors fitted to it? Almost—although we must still consider another decisive point: in the lamp, the light waves emitted by the excited photoelectrons escape too quickly to receive support from other photoelectrons, which means that the stimulated emission cannot take place, and the individual wave trains cannot have their life prolonged. The most varied waves are emitted entirely incoherently. The mirrors in the laser are designed to prevent the escape of the light waves along the axis of the laser, to provide enough time for the amplification of the light waves by stimulated emission to occur. But no mirror is perfect and therefore cannot keep the light permanently in the laser; other factors, such as scattering, also cause light to be lost. In addition, all the practical uses of the laser demand that the mirrors let some of the light escape; after all we want to illuminate all manner of objects with the light from the laser.

This makes the question of when laser light can be generated a question of quantity, for we must excite the photoelectrons in such rapid succession that they can amplify the light waves quickly enough and effectively enough to compensate for the losses caused by the mirrors. In other words, we must ensure that the energy loss of the waves is balanced by an energy gain by stimulated emission. This shows that the transition from lamp light to laser light occurs abruptly when we increase the electric current we send through the tube. There is a critical range of current intensity where a minor change changes the state of order of the laser dramatically. We can maintain the activity of the laser only by continuously supplying it with energy, for instance in the form of electric current. At the same time energy is continuously emitted in the form of laser light (and other losses). The laser then exchanges energy with its surroundings all the time; it is an open system. It also becomes a system far removed from one of thermal equilibrium like the internal combustion engine.

The sudden occurrence of a state of macro-order strongly recalls the behavior of an iron magnet or of a superconductor, which also involves states of order of entirely new physical properties, although unlike the laser these systems are in thermal equilibrium with their ambience. It therefore came as a surprise to many physicists when we were able to show in Stuttgart, together with an American group, that the laser transition exhibits all the properties of an ordinary phase transition including "critical fluctuations" and "symmetry break." The laser is thus a bridge between inanimate and animate nature. It receives its state of order through self-organization at the very moment when we increase its energy supply. Like all biological systems, the laser is an open system. Chemical lasers more than any others provide an interesting link with vital processes, because here a kind of metabolism occurs. Substances such as hydrogen and fluorine are added to the chemical laser. Both react violently with each other, creating new partnerships between the atoms of the hydrogen and the fluorine; the chemical bond between two partners is established with such force that photoelectrons are excited in the process; these emit laser light in the way we have described.

Here, then, energy is generated on the basis of chemical reactions. The chemical energy, usually liberated in the form of heat, is converted into the orderly energy of the strictly periodic wave motion of the laser light. We are faced with a metabolism as it were, in which inferior combustion energy is transformed into superior light energy. It can be compared to an engine in whose cylinder a gas mixture explodes. The thermal energy, distributed over many degrees of freedom, is converted into the kinetic energy of the piston, ultimately driving the automobile. We shall find time and again that the transformation of microenergies into the macroenergy of a few degrees of freedom appears to be one of the fundamental principles of biological processes.

We can produce laser activity not only by increasing the current intensity, which excites the individual photoelectrons more frequently. We can also envisage another process, in which we leave the pump frequency per atom unchanged but simply increase the number of laser atoms more and more, when we shall see that below a certain number of these atoms no laser activity takes place. Laser activity will, however, start quite suddenly as soon as the number of laser atoms is increased beyond this critical number. What we are in fact confronted with here is a change from quantity to quality.

These examples demonstrate that processes of self-organization can be generated in various ways. We shall have many opportunities to discuss them in the context of biological applications.

Nor is this the only link the laser provides with biology. By fitting the mirrors, we create for the laser atoms and the light waves they generate a strictly defined "environment." According to physical laws, only certain wave lengths fit between two parallel mirrors (fig. 5.10), so that only these waves can be used as laser waves. It may happen that the preferred wave that the photoelectrons of the atoms want to emit does not fit between the mirrors. But the electrons will not abandon the emission of laser light; they will choose that wave whose cycle most closely approximates that of their preferred wave (but only within certain limits). If we slowly change the distance between the mirrors, the laser light emission of the electrons changes accordingly—the electrons adapt themselves to their new environment. And now something most remarkable can happen: a new wave, more closely approximating the electrons' preferred wave than the one they had hitherto adopted and supported, may fit between the mirrors; now first a few electrons will spontaneously start, in a kind of fluctuation, to emit their energy in this new wave, before very quickly all the other electrons support this new wave and completely abandon the old one; i.e., they adapt themselves to a new "mirror environment," triggered by a fluctuation.

Both in the laser and in liquids a state of macro-order can be achieved by increased energy supply. If we increase it progressively with liquids, patterns of ever-increasing complexity will be formed until they are superseded by turbulence.

Exactly the same applies to the laser. If we increase the current intensity further, the laser will suddenly start to emit regular, incredibly short and intense flashes of light, each of which can radiate as much power as all the power stations of the United States combined. Each flash lasts one trillionth of a second. These light flashes—also called ultrashort laser pulses—are the outcome of the cooperation of many different waves. Competition between them has ceased and been replaced by a huge shared effort. Lastly, our theory predicts that lasers can produce another new kind of light: turbulent light, a new field of research for the experimental physicist.

Chapter 6

Chemical Patterns

Arranged Marriages, Chemical Style

Modern chemistry provides us with particularly beautiful examples of extensive patterns. It is general knowledge that certain chemical substances can react with each other, forming new compounds as they do. The most familiar examples are obviously combustion reactions in which chemical elements, such as carbon and oxygen, combine. Frequently a chemical reaction does not occur spontaneously, as this example shows. A minimum temperature, for instance, is needed for ignition. But the chemists have found that other possibilities often exist of initiating or at least facilitating chemical reactions. With the addition of certain substances, a chemical reaction can proceed that without the addition would take place very slowly if at all. These substances may be sheets of certain metals, for example platinum; they remain unchanged during the chemical reaction. They can be likened to marriage brokers, who introduce prospective partners to each other; i.e., they permit the chemical substances to enter into new partnerships. The chemist calls these "marriage brokers" catalysts (fig. 6.1). Here the chemists have discovered a phenomenon previously regarded as a rare curiosity but now becoming increasingly important: substances exist that are capable of catalyzing themselves. This sounds highly complicated, but it means no more than that the molecules are

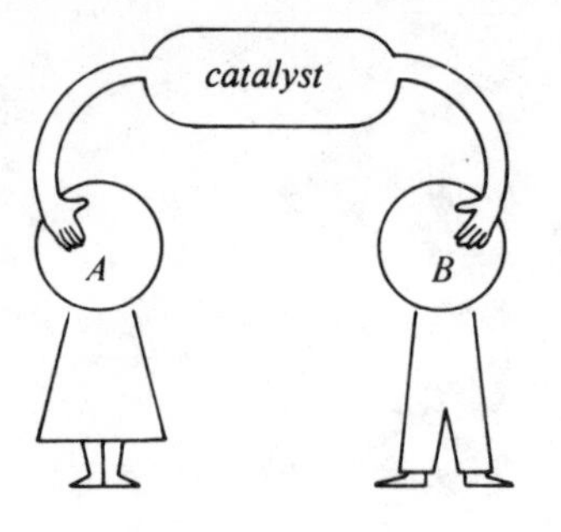

FIGURE 6.1 The catalyst as chemical "marriage broker."

able to multiply themselves, so to speak. They succeed in converting and combining other molecules so that they produce molecules of their own kind (fig. 6.2). We can already see in this capability a property of life and will not be surprised to meet this kind of reaction again in the theory of evolution. The reaction in which substances catalyse themselves is called autocatalysis. What happens in the micro-, what in the macro-range during chemical reactions? In the microrange the substances consist of individual molecules that, as we all know, are in turn made up of atoms. These molecules, let us call them type 1 and type 2, meet in a chemical reaction and form a new molecule, type 3. As they do, their physical and chemical properties, for instance their colors, change. This is clearly evident during reactions. A red liquid, for example, may thus be produced by a blue and a colorless one (fig. 6.3). Normally the new substance is dyed quite uniformly and remains permanently in this state. But this is not necessarily so, which brings us to the real subject of this chapter: during the course of this century a few highly complex reactions were discovered that produce macropatterns whose dimensions are a billion times larger than those of the molecules.

Chemical Clocks

The most famous example is a reaction scheme found by the Russian B. P. Belusov and later systematically investigated by A. M. Shabotinsky. The reaction is too complex to be described here. But the chemical patterns that it forms are of great interest. In the course of time the

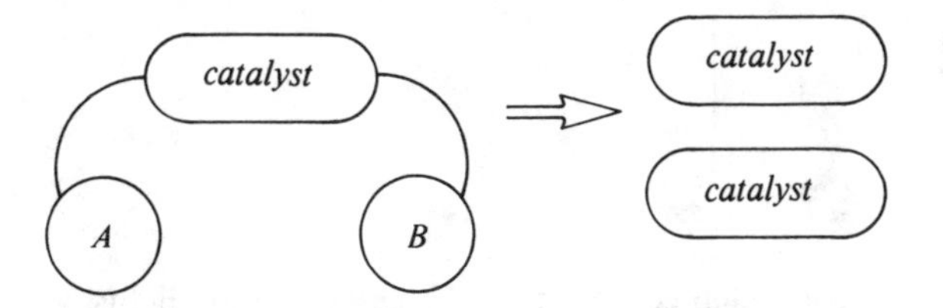

FIGURE 6.2 A catalyst combines two molecules so that a catalyst of its own type is reproduced. This is called autocatalysis.

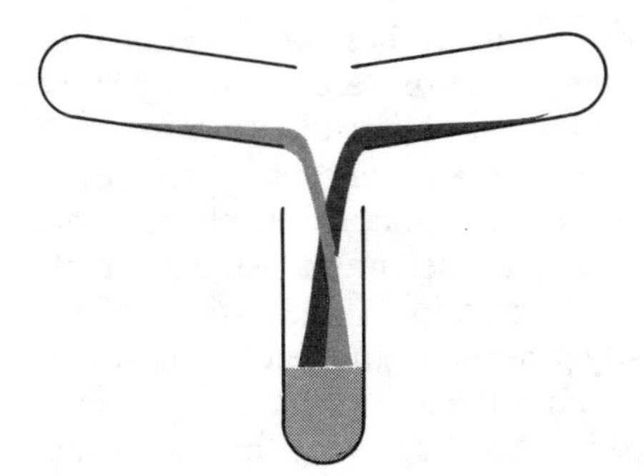

FIGURE 6.3 Pouring together and mixing of chemical substances usually leads to a homogeneous end product.

color of the liquid changes periodically from red to blue and back, and so on (fig. 6.4). A clock could be built on the basis of this reaction, for clocks are nothing but instruments that continuously indicate the duration of certain periods. We must point out that in the original experiment the substances were brought together once and for all and thoroughly mixed; the reaction, left to itself, showed the periodic color change. This, however, does not recur indefinitely; after some minutes a final state of rest will be reached.

It is possible, however, to modify the experiment by the continuous introduction of fresh substances into the reaction vessel and the removal of the metabolons. This will indeed maintain a permanent reaction of periodic color change.

The discovery of such fluctuations is of great importance to biology.

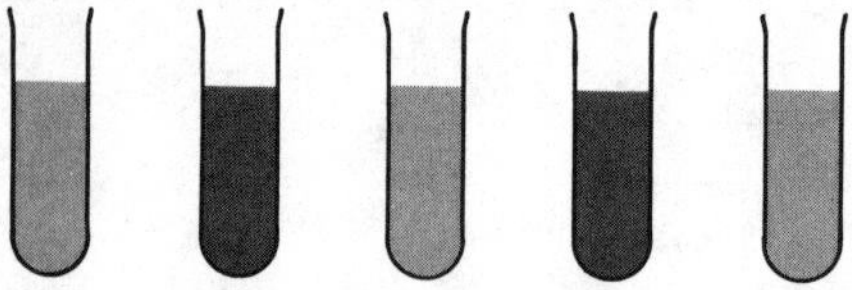

FIGURE 6.4 The periodic color change from red to blue during the Belusov-Shabotinsky reaction.

After all, the activities in living organisms are based on chemical or electrochemical processes, where many reactions are cyclic. Once we have understood the reason why chemical clocks function we shall have taken a long step forward in the knowledge of cyclic activities in the organism, such as the heartbeat. In these fluctuation phenomena, as in the laser, the concept of the order parameter and of the principle of enslavement comes into its own. At certain concentrations of the added substance the otherwise steady progress of the reaction becomes unstable and is replaced by the periodic change—i.e., a fluctuation—that plays the part of the order parameter, enslaving the individual molecules: it forces them to enter cyclically into new compounds and to leave them, etc., so that macroscopically the liquid will periodically appear red and blue. It is possible to treat such fluctuation processes mathematically and to determine the significance of the order parameter with great accuracy.

More recent research has shown that the metabolism of an individual cell linked to energy conversion is cyclic, periodic.

Chemical Waves and Spirals

Some phenomena are much more beautiful and much more complex than those just described. Several such patterns are illustrated in figure 6.5. They, too, are the result of the Belusov-Shabotinsky reaction, where in initially random centers blue dots form on a red background and grow into blue disks, in which a red dot appears and quickly grows into a red disk. Another blue dot is produced in this, and the cycle repeats itself. Concentric blue rings travel outward. In other experimental conditions, as when a nail is moved through the liquid, spirals will be produced, of which a time sequence is illustrated in figure 6.6.

It appears difficult at first glance to understand the origin of such macropatterns, but an easily constructed example comes quickly to hand. The origin of concentric rings can be compared to a prairie fire. The red background represents an area of dry grass. If we start a fire at one point in dead calm conditions it will spread evenly in all directions, in the form of a circle. If we assume blue to be the burned ground, a blue spot is produced that spreads progressively outward. In the center the grass will grow again and become dry, producing a red patch. Because the grass grows again behind the spreading front but is not yet combustible, the red patch will spread outward until the grass at the center has become dry enough to burn, and the whole cycle starts afresh. In the chemical reactions discussed no external interference in the form of ignition is necessary. The system itself is hypercritical, as it were, and can spontaneously initiate the reaction that produces the blue dots. Apart from this the phenomena are quite similar. Both with the grass and in the Belusov-Shabotinsky reaction, combustion means that certain chemical changes are taking place. But then a reaction occurs that leads to the restoration of the original state.

In the waves or spirals of the Belusov-Shabotinsky reaction the molecules reacting with each other must meet; i.e., they must be able to move. They do this through diffusion, a process familiar in everyday life. When we take up an ink blot with a piece of blotting paper, the ink will diffuse into the paper where it spreads further, forming another ink blot there. The macroprocesses described here are based on an interplay between chemical reactions on the one hand and diffusion on the other. They are therefore defined by equations, which are called reaction diffusion equations by the scientists but are outside the scope of our discussion. The only point of importance to us is that here, too, mathematical treatment proves the existence of order parameters that regulate the development of the space/time patterns. Depending on the type of order parameter, they may be circular waves or spirals.

A New Shared Principle

As we have seen in the concrete examples of laser physics, the physics of fluids, and finally of chemistry, we meet time and time again the

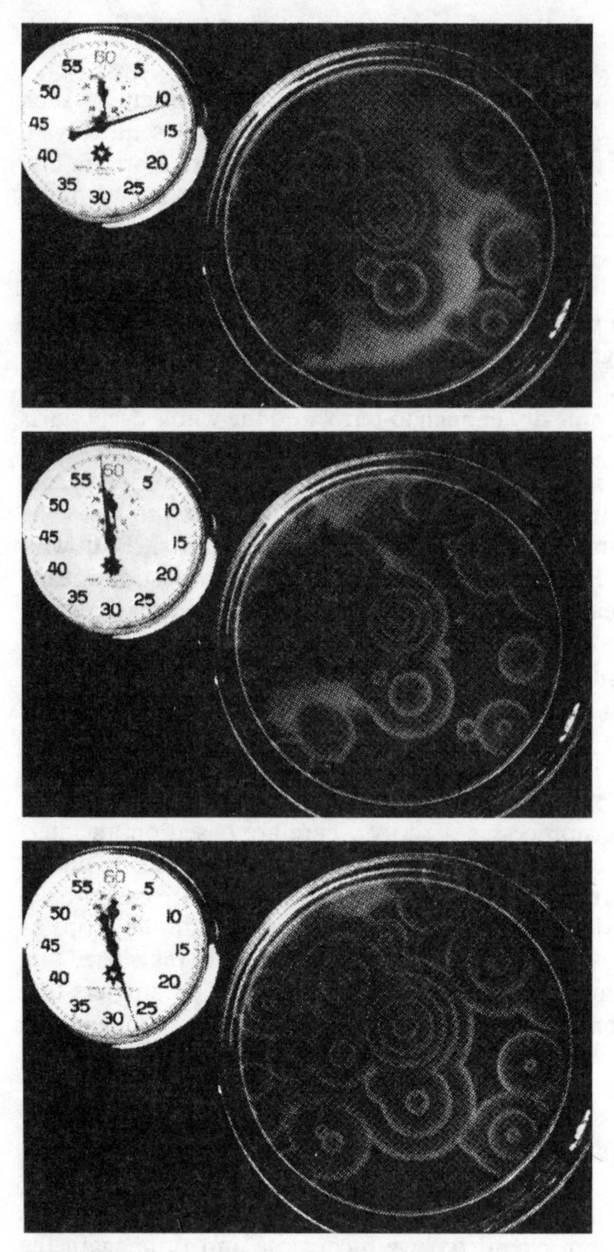

FIGURES 6.5 and 6.6 Chemical patterns in the form of concentric circles (*above*) or spirals (*opposite page*). The circles move outward; the spirals rotate.

concept of the order parameter on the one hand and of enslavement on the other. These concepts run like a thread throughout this book. In the chemical reactions we are for the first time becoming aware of a new common feature. The chemical oscillations and waves we have just described are always based on autocatalytic processes. By its presence and cooperation a certain type of molecule makes the production of more molecules of the same type possible. From this aspect the events in the laser appear in a new light. Here, too, it was an already existing light wave that by its very presence forced the electrons of the atoms to yield their energy for the amplification of this light wave. Just another autocatalytic process (fig. 6.7). The concept of autocatalysis, like those

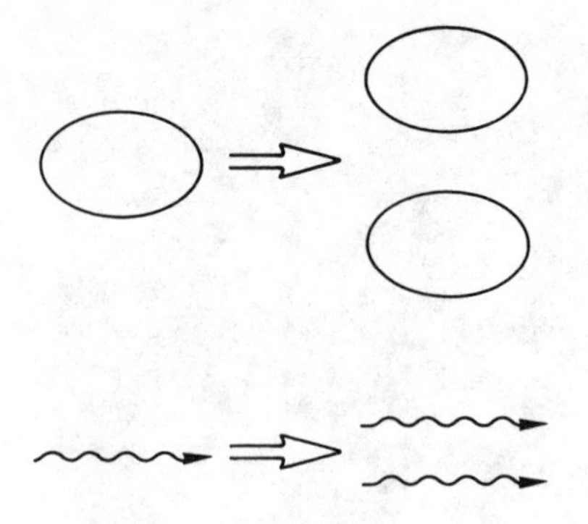

FIGURE 6.7 Analogy between autocatalysis of molecules (*top*) and amplification (= multiplication) of light waves in the laser (*bottom*).

of order parameter and enslavement, has gained an importance far beyond the field of chemistry. In this sense the rolling motion of liquids, too, has the character of autocatalysis. The developing rolling motion is amplified by the fact that initially such a motion, even a very slight one, came about by pure accident. Autocatalysis and the developing instability of forms of collective motions are one and the same phenomenon. This shows that nature obviously uses the same principles time and again to produce orderly macromotions or patterns.

Chapter 7

Biological Evolution. The Fittest Survives

As late as at the beginning of last century the origin of the various plant and animal species was to the human intellect a strictly guarded secret of nature. A decisive breakthrough was achieved by the Englishman Charles Darwin (1809–1882). During his extensive voyages of exploration to far distant countries such as South America, the enormous variety of flora and fauna and the ingenious organs they possess to ensure survival attracted his attention. After many years of deliberation, this led him to formulate entirely new theories about the origin of species; the basic postulates are fully accepted to this day. We call it Darwin's theory or Darwinism, forgetting that quite independently of Darwin but at the same time a young Englishman, Alfred Russel Wallace (1823–1913), had exactly the same ideas.

In 1856, two years before he received the, to him, shocking communication from Wallace that contained the latter's own formulation of the theory of evolution, Darwin wrote his now famous letter to Charles Lyell (1797–1875); he indicated that he was not quite ready to publish his views, contrary to Lyell's suggestion that publication would serve to prevent anybody else from anticipating them. Darwin said: "I do not like the idea of publishing on account of priority, but I should be annoyed if someone were to publish my ideas before me." (Goethe's saying—"Two souls, alas, dwell in my breast"—seems typical of many scientists' attitude of which Darwin was one example among many others quoted by the sociologist R. K. Merton.)

In 1858 the blow fell. What Lyell had warned him of and Darwin had refused to believe, happened. He wrote to Lyell about this shattering event: "[Wallace] today sent me the enclosed and asked me to pass it on to you. It appears to me to be well worth reading. Your words have come true with a vengeance—that someone might anticipate me. I have never seen a more striking example of coincidence. If Wallace had copied my manuscript draft of 1842 he could not have made a better brief summary. Even his expressions now stand as the headings of my chapters. Thus all my originality, whatever it may mean, will be destroyed."

Modesty and disinterestedness urged Darwin to give up his claim to priority, but his desire for acknowledgment and authorship urged him to consider that not all was lost. At first, with characteristic generosity but without pretending equanimity, he made the despairing decision to step aside completely. A week later he again wrote to Lyell, suggesting that he could perhaps publish a short version of his long-existing text, probably a dozen pages. Yet he wrote in distress: "I cannot persuade myself that I can honorably do this." Made distraught by mixed feelings he concluded his letter: "My good, dear friend, forgive me. This is a futile letter, influenced by futile feelings." In the endeavor to cleanse himself finally of these feelings, he added a postscript: "I shall trouble neither you nor Hooker ever again in this matter."

On the following day he again wrote to Lyell, this time to revoke his postscript, but still torn. As fate would decree, at this very moment Darwin was shocked by the death of his young daughter. He brought himself to complying with the wish of his friend Joseph Dalton Hooker (1817–1911), and sent both Wallace's manuscript and his own original version of 1844. He wrote: "Solely that you can recognize from my own handwriting that you have read it. Do not waste much time. I feel unhappy that I worry so much about priority."

Other members of the scientific community did what the tortured Darwin was not prepared to do for himself. Lyell and Hooker took the matter in hand and arranged the momentous meeting of the Linnean Society at which both papers were read.

This was the official hour of the birth of the theory of evolution. By a hair's breadth we would have spoken of Wallacism instead of Darwinism today. In Chapter 16 we shall discuss the question of why one became famous while the other is now all but forgotten.

Here, then, are Darwin's fundamental theses: he claims that a development takes place in nature in which complex creatures evolve from simpler ones. The basic role is played by the interaction between the hereditary traits—i.e., the genotype—on the one hand, and the individual animals or plants that appear to us as such in their developed form—i.e., as the phenotype—on the other. Darwin assumed that the hereditary traits can change spontaneously. These changes are the mutations. As we now know such mutations can be demonstrated in the genes, which transmit the hereditary traits; they are, of course, microscopic.

Because of the changed hereditary traits the properties of animals and plants also change. The offspring of white butterflies, for instance, can acquire black wings, limbs can become crippled or appear in a changed shape. These mutations may enable the animal to make better use or condemn it to make poorer use of its environment. For example, with a changed bill birds may be able to eat insects that they previously were unable to pick up. Nature never ceases to surprise us with the abundance of its most varied forms, which very often quickly reveal that they are highly functional. This functionality was felt to be teleological in earlier centuries; i.e., God created the animals so that they can find their food most efficiently. But according to Darwin these forms are products of the accident of mutation on the one hand, and of selection on the other. The various animals, more or less well adapted to their environment, compete for their food. Other forms of competition are also possible, with birds for instance the search for nesting sites or for protection against hazards. Rivalry among the various species begins and only the more efficient one survives. These, then, are the fundamental theses of Darwinism.

Here, however, we are confronted by a number of difficulties that are recognized mainly by biologists and natural philosophers. To begin with, the claim that "the fittest survives" can be likened to a cat biting its own tail. He who survives is defined as the best. But we can cut this Gordian knot by citing an example from the inanimate world, because Darwinism covers both animate nature and inanimate matter. The laser has already given us an instance where we found competition among laser waves of which also only one survives. This we can obviously define as the "best." The important point, however, is that in laser physics we can calculate from the outset which mode or which wave will survive—i.e., which is the best. There are objective criteria that enable us to predict the winner in advance of the entire process, with a certain proviso: occasionally several waves will be candidates simultaneously. The symmetry between them can be broken, or in other words the final choice between them can be made only through an accidental oscillation, which we cannot predict. But in addition to these distinctive modes there are many, many others of which we can say with absolute certainty that they will not survive. Laser dynamics thus present a physical model that allows us to copy the evidence of Darwinism experimentally as well as mathematically. Here the claims of Darwinism are very quickly verified in all their harshness. The process of the laser oscillations which "live," as it were, on the excited atoms, can be directly transposed into animate nature: several species that live on the same food will compete, and as a result only the most efficient species, say the one able to take up the food most quickly, will survive.

Competition among the Biomolecules

Such an analogy between selection processes in animate and in inanimate nature is not confined to the laser oscillations. Another bridge between "inanimate" and "animate" is established in a closely related sense by Manfred Eigen's theory of evolution. This is based on the claim that the hereditary traits are transmitted by certain "bio" molecules, which we shall discuss in greater detail in Chapter 9. The only important point in the present context is their ability to multiply through autocatalysis, like laser modes, and, also like these, entering into competition. In the original version of Eigen's theory the equations that described the multiplication of the biomolecules had exactly the same form as those that dealt with the "multiplication" of the laser waves. It can hardly be a coincidence that such an agreement should exist between two entirely different fields; in fact it points to the existence of generally valid principles, which we meet indeed time and time again in this book.

The particular attraction of this theory of evolution is of course that it forges a link between inanimate and animate nature through mutation and selection and thus through a "higher development" of the biomolecules, and reveals a more or less continuous transition from "inanimate" to "animate." Without doubt much research remains to be done, especially in the biochemical area, but a promising start has been made.

To complete the picture, Manfred Eigen and Peter Schuster refined the ideas about the autocatalytic multiplication of the biomolecules within the last few years.

In the simplest case there are two types of molecule, A and B. Each of them multiplies by autocatalysis. In addition, type A assists type B with multiplication as a catalyst, and type B in turn is a catalyst that assists type A with multiplication. The scheme is illustrated in figure 7.1 left, and can be expanded to include several types of molecule. Three types, A, B, C, multiply autocatalytically; furthermore, A helps with the multiplication of B, B with that of C, and C with that of A, each in the role of catalyst. Eigen and Schuster call such smaller or larger cycles hypercycles. In turn, the hypercycles may suffer mutations as well as compete with each other. Darwinism is at work in laser waves, in biomolecules, in hypercycles, and in the animal and vegetable kingdoms.

The fact that Darwin's rules govern both animate and inanimate matter is a pointer toward their enormous significance. They are also of direct importance to sociology, which has to tackle such problems as

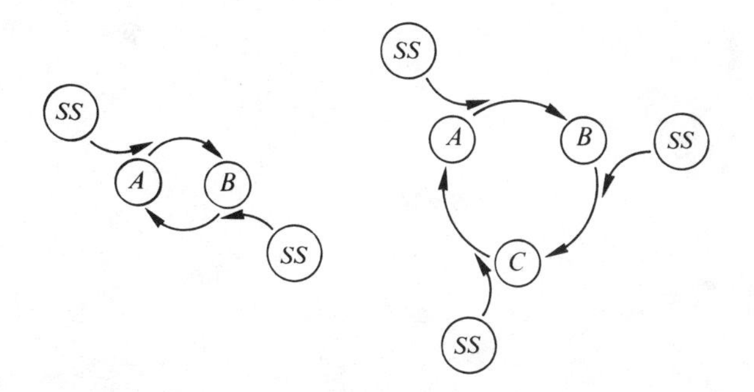

FIGURE 7.1 Two examples of Eigen's hypercycle. *Left:* The molecule type A increases autocatalytically, which, however, calls for the participation of the molecule type B as catalyst. Conversely the molecule type B increases autocatalytically, which requires the participation of the molecule type A as catalyst. "SS" indicates that these molecules are produced from certain simple substances. *Right:* Hypercycle with three different types of molecule, each of which increases autocatalytically but still requires the participation of the other types of molecules, as indicated in the diagram. The cycle of the participating molecules can be much larger.

professional and economic rivalry. Applied in this sense, Darwin's rules would result in firms that make the same product but sell it at different prices; they are subject to selection in the marketplace, until as a result of competition only one is left to monopolize the market. Is this forceful trend, which ultimately must produce a giant combine, indeed natural—i.e., should only the absolutely most fit be the victor in the harsh struggle for survival? Here, too, nature has shown us some loopholes, as we shall explain in the next chapter.

Chapter 8

To Survive without Being the Fittest: Specialize, and Create Your Own Ecological Niche

On close inspection the thesis that only the fittest survives presents a number of profound problems. On the basis of this thesis it is surprising that there should be such an abundance of different species in the world. Are all of them the fittest? This is reason enough for us to scrutinize the question of survival in some greater depth.

Nature has indeed contrived innumerable tricks to outsmart the theory. To begin with, obviously competition can develop only if the various competing species live in the same area. There is clearly no competition between land animals living on continents separated by the sea. In Australia therefore a fauna quite different from that of other countries has evolved, such as the marsupials, of which the kangaroo is only one example. But even when the species live close together they have often been successful in establishing new environments. Birds, for example, have opened up different sources of food for themselves by developing different forms of bill (fig. 8.1). These species have thus evaded the need for a hard struggle for survival by creating an ecological niche for themselves. To this extent it can of course be claimed that they have become the fittest in their special field, because they are the only species to have acquired this particular special ability. An ecological niche is, as it were, a wildlife reserve, a protective zone in which a certain species can live on its own without outside interference. Our example of the food sources shows that ecological niches

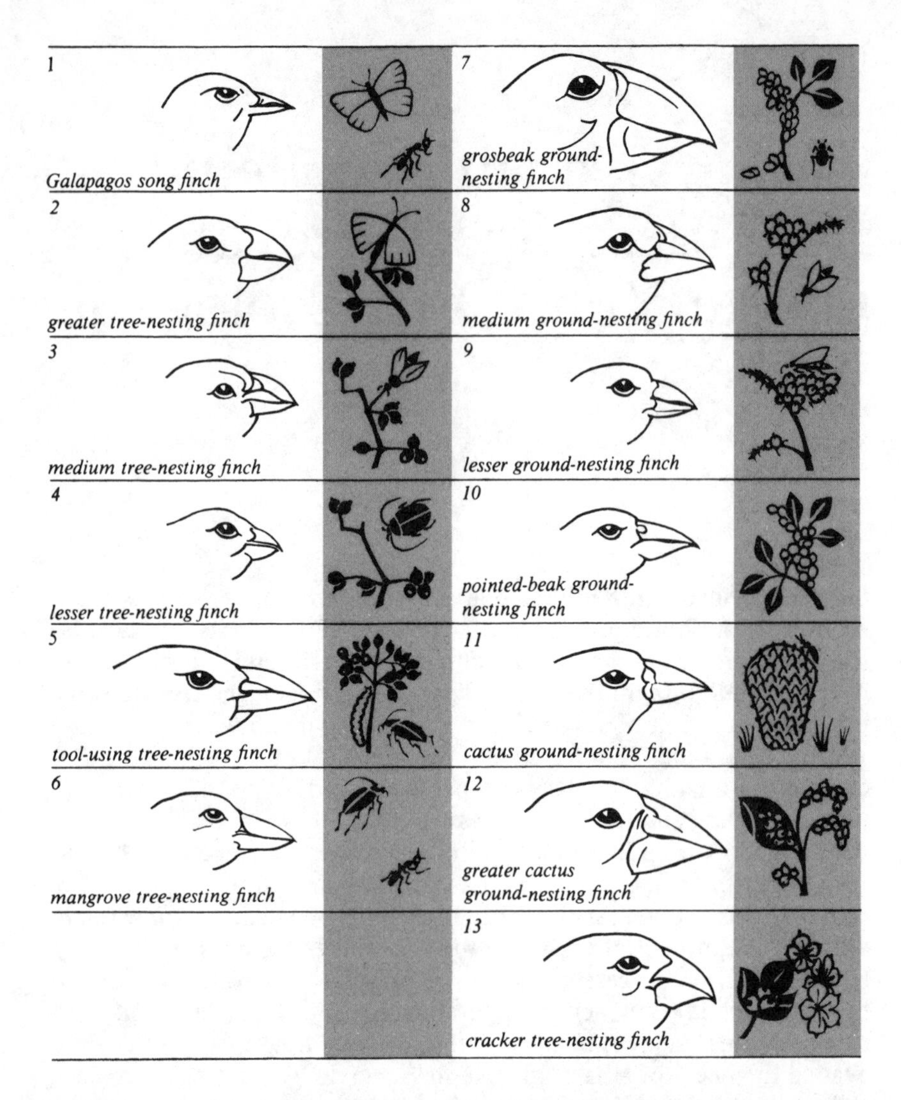

FIGURE 8.1 Examples of various types of bird beaks (of finches), showing their high degree of specialization for certain uses. The food taken by each species is shown symbolically. Finches on an island cut off from the rest of the world have produced widely differing forms of beak. New races have resulted, but not a new family—they have always remained finches. Darwin described them first; Lorenz drew the pictures.
1 = insectivores; 2-6 = mainly insectivores; 7-12 = mainly herbivores; 13 = herbivores.

need not consist only of territories isolated from each other, although of course territorial separation functions even more effectively as an ecological niche.

Coexistence through specialization is by no means confined to animate nature. In the laser, different light waves may occur simultaneously and not compete with each other provided they derive their energy from different atoms. In professional and economic life, the question of competition also plays a decisive part. We shall return to it in due course.

Interestingly, nature offers us examples not only of survival through specialization but also through generalization, when for instance a species of animal, such as the wild pig, can live on a very broad spectrum of food.

An example of particular interest to the survival in the harsh struggle for life is symbiosis, in which widely different species assist each other, are even essential to each other's existence. Nature offers us a wide range of examples. Bees, which live on the nectar of the flowers, at the same time ensure fertilization and therefore an increase in the providers of their food, birds that fly into the gaping mouth of crocodiles to "clean" the reptiles' teeth, ants that keep greenflies as "milch cows." It was believed that the Tree of Calvary, off which the dodo lived, was doomed to extinction because only the dodo could process its seeds by digestion so that it germinated; and the dodo died out. (According to recent reports, biologists have discovered that turkeys can process the seeds of this tree, which can grow to an age of several hundred years.)

But we must not lose sight of the overall picture with all these detailed features. Usually it is by no means just two or three animal species that compete or live in symbiosis with each other. In fact, natural processes are infinitely interlinked. Nature is a highly complex synergetic system.

It is a question of fundamental importance whether the various interlinked natural processes can produce an equilibrium. "Ecological balance" has become a very popular subject of late, a state which is more and more upset by human interference. But the most recent research findings strongly suggest that even without human interference the ecological or biological balance is by no means as perfect as has been assumed for a long time.

We think of balance generally in terms of the static kind in which the number of, say, a certain species of bird remains practically unchanged through time. This is, however, not always so in nature. Such changes can be caused by natural catastrophes, a fact with which we are all familiar. A winter may be too hard, a summer too dry and hot; a frost may kill the blossoms, and the bees cannot find enough food. Other examples of disturbed balance come about with the occurrence of pests, for instance of mice or of May bugs, which will have a devastating effect on other spheres of life.

But let us disregard the possibility of such imbalance. Even after such a natural catastrophe we take it for granted that the previous balance will be restored. As we shall see later, the theory of the free market economy is based on an equivalent assumption.

But is nature really as stable as this? The number of examples is indeed on the increase that the balances in nature are by no means all static. At the beginning of the twentieth century, fishermen in the Adriatic noticed that their catches fluctuated rhythmically. As they very quickly discovered, this was due to a rhythmic fluctuation of the fish population (fig. 8.2). In the twenties, two eminent mathematicians, A. J. Lotka and V. Volterra (1860–1940), succeeded independently of each other in providing a mathematical explanation of this finding. It was discovered that the phenomenon was brought about by two species of fish, one predatory, the other prey eaten by the former. The mechanism of the cyclic fluctuation is as follows: at first there are relatively few predators, and the prey can multiply without hindrance. This leads to a situation in which the predators find more prey and are therefore able to increase in numbers. Ultimately they become so numerous that they radically reduce the numbers of their prey; this in turn leads to a reduction in the number of predators, and the cycle begins anew.

It may happen in the mathematical model that the predators by accident eat all the prey and are themselves doomed to extinction. Nature prevents this process by providing refuges for the prey, where they are safe from the predators.

A similar cycle has been discovered in Canada, where snow hares are eaten by lynxes (fig. 8.3).

Because birth and death rates are also subject to other influences, the model concepts described here are sometimes subject to criticism. But they do show that in nature static balances are not necessarily to be taken for granted.

These conditions are even more marked with some insect populations, whose numbers fluctuate quite irregularly. As we shall discuss

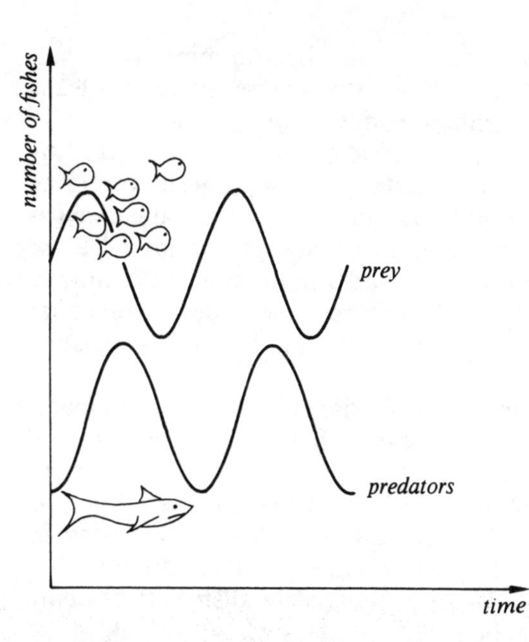

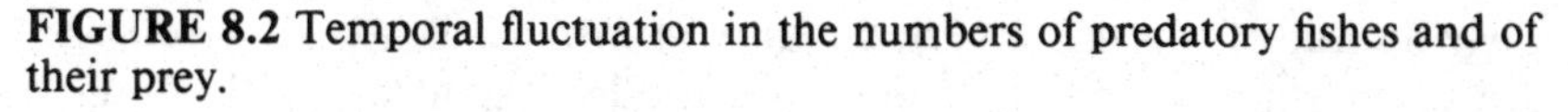

FIGURE 8.2 Temporal fluctuation in the numbers of predatory fishes and of their prey.

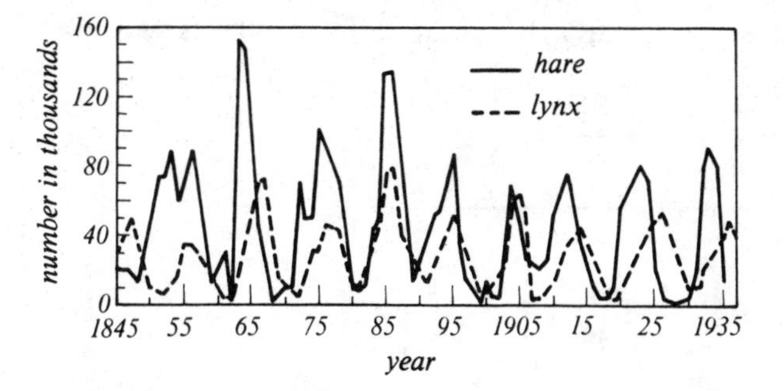

FIGURE 8.3 Periodic fluctuation in the populations of snow hares and lynxes.

later, mathematical models have been worked out which enable us to understand in depth even processes that seem to be entirely irregular (fig. 8.4).

The examples given here show the idea of a static biological balance to be too naive. On the other hand we must bear in mind that a balance, having really established itself, is extremely sensitive; this leads us back to one of the fundamental theories of synergetics.

A number of examples from physics and chemistry in the preceding chapters have demonstrated that at certain critical points even minute changes in environmental conditions can produce dramatic changes on the macro level. In these examples we always considered those changes that achieve a higher degree of order in the systems. Naturally all these processes can also be looked at from the opposite direction; i.e., minor changes in the environment can destroy an existing order.

Many events affecting animal populations can be described in mathematical terms so that the mathematical methods of synergetics can be applied to these problems. Far-reaching analogies between occurrences in animate and in inanimate nature can in turn be established on this mathematical level. The results can be presented in a few words: in animate nature, too, even minor changes in external conditions often create entirely new states of order—i.e., entirely new distributions of the various species. We already have proof of this when we look at the distribution of the most varied plants in mountain areas. Here there are often sharply defined zones of altitude that form the boundary between different belts of plant life, quite similar to the earth's climatic belts. This example shows clearly that quite different plants will pre-

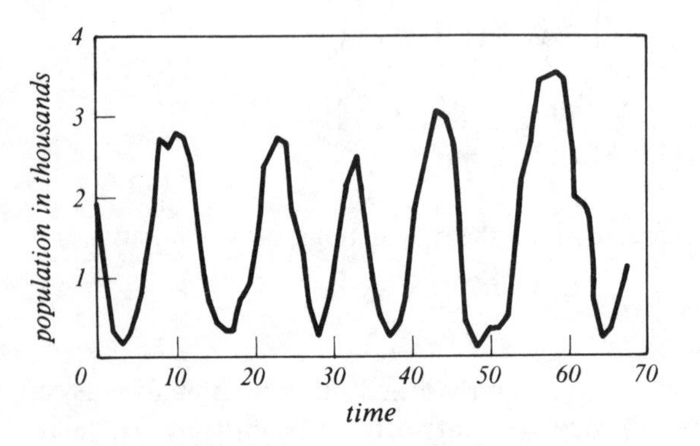

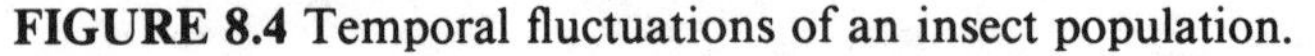
FIGURE 8.4 Temporal fluctuations of an insect population.

vail over others quite abruptly even after only a very minute change in the mean annual temperature. We must expect exactly the same kind of change when we alter the environment by artificial interference. If we pour effluents into a river, it would be naive to anticipate a 10 percent decrease in the fish population as a result of a 10 percent increase in pollution. In fact, at a critical point even a minor increase in pollution may lead to the complete extinction of the fish population; in other words, the balance of the water may suddenly shift. Here a basic principle of synergetics, which we meet time and time again, is more obvious than ever: at certain points of instability even minor changes in the environment may result in quite dramatic changes in the entire system.

In conclusion let us once again turn to nature. Here, too, environmental conditions change, for instance when the climate changes. It should be clear by now that even a negligible change will be able to produce basically new processes of selection, and thereby to advance development.

But "advancing a development" does not imply that the newly developing species will be objectively better than those they displace. The new species are merely better adapted to the new conditions of survival. This may even involve changes that can be thought of as retrograde. Complex creatures may be replaced by others of simpler design. This can happen even on the level of the biomolecules, which in changed environmental conditions may shed part of their hereditary traits because they can manage and even procreate more quickly without them. Such experiments were conducted by Sol Spiegelmann on the biomolecular DNA of certain phages.

In animate nature objects are involved that are entirely different from those in the processes of chemical reaction, those in the laser or in a liquid, but the same basic principles are at work in all of them. Here the species function as order parameters; they may compete, cooperate, or coexist.

Minor changes in environmental conditions can bring entirely new order parameters or systems of order parameters to bear, provided of course that a new order parameter, here a new species, is created in the first place. In the laser this was the spontaneous generation of a light wave, in the liquid motion a minor thermal variation, in the chemical reaction an initial reaction or the spontaneous generation of a molecule of a new kind. Here again the interplay between accident and necessity becomes clearly apparent. Changes in the environment create conditions in which new states of order, described by the associated order parameter, can assert themselves. But first there must be the accidental creation of a new species, in biology by mutation. Alternatively a species that had existed hitherto in small numbers (for instance, in an ecological niche) may explosively multiply and become dominant.

As in all the previous cases, a peculiar relation exists between order parameter and individual, which the following example will illustrate. In a number of instances a simple mathematical magnitude can be correlated with the order parameter—i.e., with the number of individuals of a species. The temporal change of this number can be obtained let us say through counting; sometimes the change can be predicted. Innumerable individual fates are covered by these quantitative details, determined by the order parameter, the overall population, only in an overall way but with inexorable harshness. If less food than is necessary for the existence of the individual is available in an underdeveloped country for a certain period, the number of individuals, their order parameter, must decrease; who meets this cruel fate is neither here nor there. We have a similar situation in economic life, for instance with unemployment, or in the state. Generally, statements can be made only about the order parameters, not about indirectly affected fates, a point to which we shall return presently.

Chapter 9

How Do Biological Organisms Originate? Heredity through Molecules

We must now progress from animate nature in general, from the dynamics of the cooperation of the most varied creatures, to the individual creatures. Whereas the various species surprise us with the great variety of forms, members of any one species are characterized by the constancy of the form in which they are reproduced in the long term. The creation of forms, then, must be subject to strict rules. But how can forms originate in the first place and their creation be regulated? The simplest answer would be to point to heredity. We know, after all, that physical and without doubt also mental traits are inherited through a substantive carrier—i.e., a chemical compound specific to each species. The chemists have the complicated name of desoxyribonucleic acid, DNA, for it. DNA consists of two helical, interlaced chains of molecules, also called double helix (fig. 9.1). Like a string of pearls containing four different types of pearl, generally four different molecules are strung, seemingly at random (fig. 9.2), on one of the strings. These four molecules have names incomprehensible to the layman: A (adenine), C (cytosine), G (guanine), T (thymine). As long as we remember the initials, we may forget the names. If we gave the various molecules different colors, they would make a very colorful pearl necklace.

The DNA is copied in a cell, similarly to a positive picture from a photographic negative. This chemically produces ribonucleic acid,

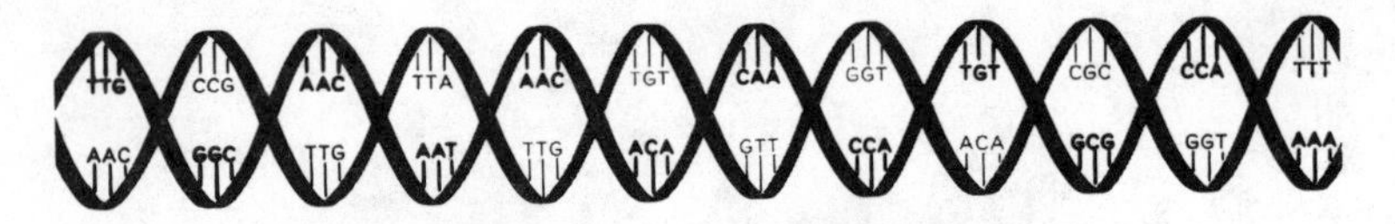

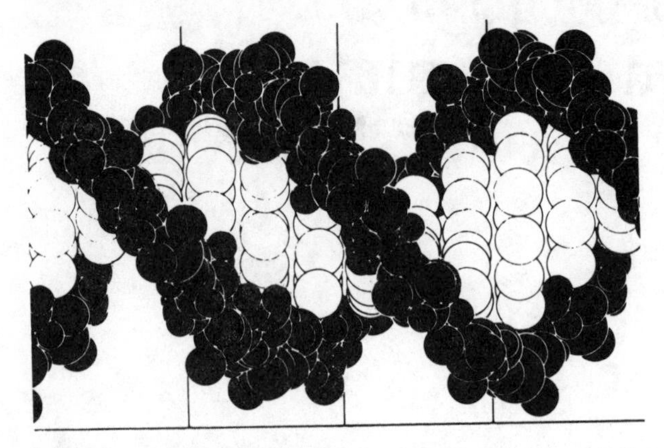

FIGURE 9.1 Molecular chains of DNA arranged in a double helix. *Top:* Longitudinal section. *Bottom:* Perspective representation.

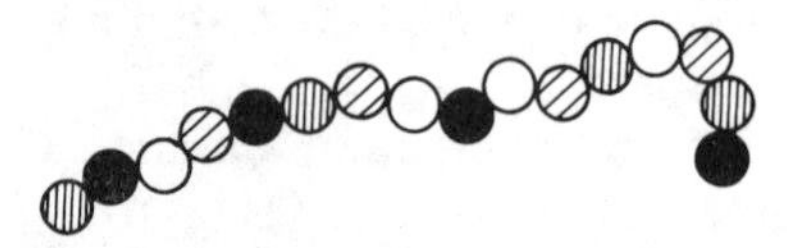

FIGURE 9.2 Different molecules are arranged, like pearls on a string, along a molecular chain.

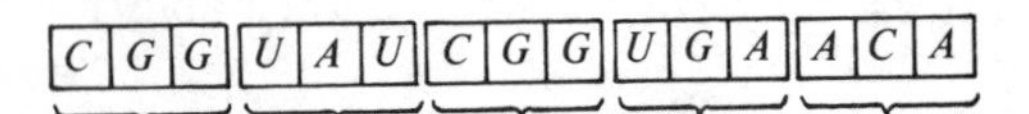

FIGURE 9.3 Example of codons containing three molecules each.

RNA. Each of the individual molecules A, C, G, T is copied to form a new molecule:

DNA	RNA
A	U (uracil)
C	G
G	C
T	A

It has been found that these individual molecules (pearls) become ordered in regular groups of three, for example

GAU, CCU, GCU, UUU,

when they form a keyword, a code for the incorporation of a certain "amino acid" (fig. 9.3).

The sequence GAU-CCU-GCU-UUU occurring in a certain RNA is thus the written instruction to the cell: build an albumin substance, a protein, placing aspartic acid first, alanine second, etc. RNA thus in general controls the material structure of the cell; to describe many important details would far exceed the framework of this book. Each triad of A, C, G, U thus is an information unit, a code word or codon, as it is also called. Depending on the creature, the DNA and RNA respectively contains from a few dozen to many millions of such codons, enough to be the page of a book or, like the human DNA, even an entire book (fig. 9.4).

The idea suggests itself that the DNA transfers the instructions, the plan as it were, from one organism to the next. To use another simile, it is like a recording tape that contains a tune.

But if we consider this idea of heredity in greater detail, certain difficulties will arise. If a plan is to be put into practice, precise instructions are required, which should be included in the plan; they should indicate, for instance, where a cell of the developing organism should be located and what its properties should be. But if we count the numbers of instructions or, to use the experts' language, estimate the measure of information necessary to build up the organism, we quickly arrive at a figure very much larger than could ever be stored in a DNA. Or, to return to the comparison of DNA with a book, a huge library would be required for the human body, for example. Nature, then, must have developed methods of managing with far fewer items of

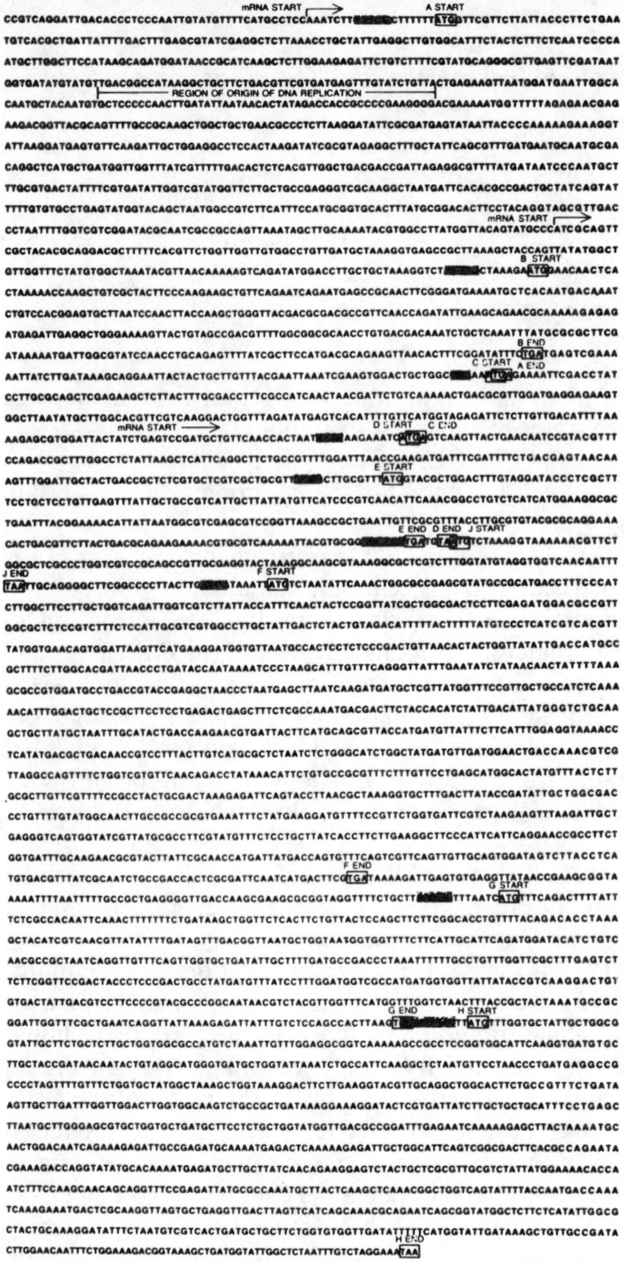

mRNA START → A START
CCGTCAGGATTGACACCCTCCCAATTGTATGTTTTCATGCCTCCAAATCTT[illegible]CTTTTTTATGGTTCGTTCTTATTACCCTTCTGAA
TGTCACGCTGATTATTTTGACTTTGAGCGTATCGAGGCTCTTAAACCTGCTATTGAGGCTTGTGGCATTTCTACTCTTTCTCAATCCCCA
ATGCTTGGCTTCCATAAGCAGATGGATAACCGCATCAAGCTCTTGGAAGAGATTCTGTCTTTTCGTATGCAGGGCGTTGAGTTCGATAAT
GGTGATATGTATGTTGACGGCCATAAGGCTGCTTCTGACGTTCGTGATGAGTTTGTATCTGTTACTGAGAAGTTAATGGATGAATTGGCA
REGION OF ORIGIN OF DNA REPLICATION
CAATGCTACAATGTGCTCCCCCAACTTGATATTAATAACACTATAGACCACCGCCCCGAAGGGGACGAAAAATGGTTTTTAGAGAACGAG
AAGACGGTTACGCAGTTTTGCCGCAAGCTGGCTGCTGAACGCCCTCTTAAGGATATTCGCGATGAGTATAATTACCCCAAAAAGAAAGGT
ATTAAGGATGAGTGTTCAAGATTGCTGGAGGCCTCCACTAAGATATCGCGTAGAGGCTTTGCTATTCAGCGTTTGATGAATGCAATGCGA
CAGGCTCATGCTGATGGTTGGTTTATCGTTTTTGACACTCTCACGTTGGCTGACGACCGATTAGAGGCGTTTTATGATAATCCCAATGCT
TTGCGTGACTATTTTCGTGATATTGGTCGTATGGTTCTTGCTGCCGAGGGTCGCAAGGCTAATGATTCACACGCCGACTGCTATCAGTAT
TTTTGTGTGCCTGAGTATGGTACAGCTAATGGCCGTCTTCATTTCCATGCGGTGCACTTTATGCGGACACTTCCTACAGGTAGCGTTGAC
mRNA START →
CCTAATTTTGGTCGTCGGATACGCAATCGCCGCCAGTTAAATAGCTTGCAAAATACGTGGCCTTATGGTTACAGTATGCCCATCGCAGTT
CGCTACACGCAGGACGCTTTTTCACGTTCTGGTTGGTTGTGGCCTGTTGATGCTAAAGGTGAGCCGCTTAAAGCTACCAGTTATATGGCT
B START
GTTGGTTTCTATGTGGCTAAATACGTTAACAAAAAGTCAGATATGGACCTTGCTGCTAAAGGTCT[illegible]CTAAAGAATGGAACAACTCA
CTAAAAACCAAGCTGTCGCTACTTCCCAAGAAGCTGTTCAGAATCAGAATGAGCCGCAACTTCGGGATGAAAATGCTCACAATGACAAAT
CTGTCCACGGAGTGCTTAATCCAACTTACCAAGCTGGGTTACGACGCGACGCCGTTCAACCAGATATTGAAGCAGAACGCAAAAAGAGAG
ATGAGATTGAGGCTGGGAAAAGTTACTGTAGCCGACGTTTTGGCGGCGCAACCTGTGACGACAAATCTGCTCAAATTTATGCGCGCTTCG
B END
ATAAAAATGATTGGCGTATCCAACCTGCAGAGTTTTATCGCTTCCATGACGCAGAAGTTAACACTTTCGGATATTTCTGATGAGTCGAAA
C START A END
AATTATCTTGATAAAGCAGGAATTACTACTGCTTGTTTACGAATTAAATCGAAGTGGACTGCTGGC[illegible]GAAAATTCGACCTAT
CCTTGCGCAGCTCGAGAAGCTCTTACTTTGCGACCTTTCGCCATCAACTAACGATTCTGTCAAAAACTGACGCGTTGGATGAGGAGAAGT
GGCTTAATATGCTTGGCACGTTCGTCAAGGACTGGTTTAGATATGAGTCACATTTTGTTCATGGTAGAGATTCTCTTGTTGACATTTTAA
mRNA START → D START C END
AAGAGCGTGGATTACTATCTGAGTCCGATGCTGTTCAACCACTAAT[illegible]AAGAAATCATGAGTCAAGTTACTGAACAATCCGTACGTTT
CCAGACCGCTTTGGCCTCTATTAAGCTCATTCAGGCTTCTGCCGTTTTGGATTTAACCGAAGATGATTTCGATTTTCTGACGAGTAACAA
E START
AGTTTGGATTGCTACTGACCGCTCTCGTGCTCGTCGCTGCGTT[illegible]CTTGCGTTTATGGTACGCTGGACTTTGTAGGATACCCTCGCTT
TCCTGCTCCTGTTGAGTTTATTGCTGCCGTCATTGCTTATTATGTTCATCCCGTCAACATTCAAACGGCCTGTCTCATCATGGAAGGCGC
TGAATTTACGGAAAACATTATTAATGGCGTCGAGCGTCCGGTTAAAGCCGCTGAATTGTTCGCGTTTACCTTGCGTGTACGCGCAGGAAA
E END D END J START
CACTGACGTTCTTACTGACGCAGAAGAAAACGTGCGTCAAAAATTACGTGCGG[illegible]TGATG[illegible]TGTCTAAAGGTAAAAAACGTTCT
GGCGCTCGCCCTGGTCGTCCGCAGCCGTTGCGAGGTACTAAAGGCAAGCGTAAAGGCGCTCGTCTTTGGTATGTAGGTGGTCAACAATTT
J END F START
TAATTGCAGGGGCTTCGGCCCCTTACTTG[illegible]TAAATTATGTCTAATATTCAAACTGGCGCCGAGCGTATGCCGCATGACCTTTCCCAT
CTTGGCTTCCTTGCTGGTCAGATTGGTCGTCTTATTACCATTTCAACTACTCCGGTTATCGCTGGCGACTCCTTCGAGATGGACGCCGTT
GGCGCTCTCCGTCTTTCTCCATTGCGTCGTGGCCTTGCTATTGACTCTACTGTAGACATTTTTACTTTTTATGTCCCTCATCGTCACGTT
TATGGTGAACAGTGGATTAAGTTCATGAAGGATGGTGTTAATGCCACTCCTCTCCCGACTGTTAACACTACTGGTTATATTGACCATGCC
GCTTTTCTTGGCACGATTAACCCTGATACCAATAAAATCCCTAAGCATTTGTTTCAGGGTTATTTGAATATCTATAACAACTATTTTAAA
GCGCCGTGGATGCCTGACCGTACCGAGGCTAACCCTAATGAGCTTAATCAAGATGATGCTCGTTATGGTTTCCGTTGCTGCCATCTCAAA
AACATTTGGACTGCTCCGCTTCCTCCTGAGACTGAGCTTTCTCGCCAAATGACGACTTCTACCACATCTATTGACATTATGGGTCTGCAA
GCTGCTTATGCTAATTTGCATACTGACCAAGAACGTGATTACTTCATGCAGCGTTACCATGATGTTATTTCTTCATTTGGAGGTAAAACC
TCATATGACGCTGACAACCGTCCTTTACTTGTCATGCGCTCTAATCTCTGGGCATCTGGCTATGATGTTGATGGAACTGACCAAACGTCG
TTAGGCCAGTTTTCTGGTCGTGTTCAACAGACCTATAAACATTCTGTGCCGCGTTTCTTTGTTCCTGAGCATGGCACTATGTTTACTCTT
GCGCTTGTTCGTTTTCCGCCTACTGCGACTAAAGAGATTCAGTACCTTAACGCTAAAGGTGCTTTGACTTATACCGATATTGCTGGCGAC
CCTGTTTTGTATGGCAACTTGCCGCCGCGTGAAATTTCTATGAAGGATGTTTTCCGTTCTGGTGATTCGTCTAAGAAGTTTAAGATTGCT
GAGGGTCAGTGGTATCGTTATGCGCCTTCGTATGTTTCTCCTGCTTATCACCTTCTTGAAGGCTTCCCATTCATTCAGGAACCGCCTTCT
GGTGATTTGCAAGAACGCGTACTTATTCGCAACCATGATTATGACCAGTGTTTCAGTCGTTCAGTTGTTGCAGTGGATAGTCTTACCTCA
F END
TGTGACGTTTATCGCAATCTGCCGACCACTCGCGATTCAATCATGACTTCGTGATAAAAGATTGAGTGTGAGGTTATAACCGAAGCGGTA
G START
AAAATTTTAATTTTTGCCGCTGAGGGGTTGACCAAGCGAAGCGCGGTAGGTTTTCTGCTT[illegible]TTTAATCATGTTTCAGACTTTTATT
TCTCGCCACAATTCAAACTTTTTTTCTGATAAGCTGGTTCTCACTTCTGTTACTCCAGCTTCTTCGGCACCTGTTTTACAGACACCTAAA
GCTACATCGTCAACGTTATATTTTGATAGTTTGACGGTTAATGCTGGTAATGGTGGTTTTCTTCATTGCATTCAGATGGATACATCTGTC
AACGCCGCTAATCAGGTTGTTTCAGTTGGTGCTGATATTGCTTTTGATGCCGACCCTAAATTTTTTGCCTGTTTGGTTCGCTTTGAGTCT
TCTTCGGTTCCGACTACCCTCCCGACTGCCTATGATGTTTATCCTTTGGATGGTCGCCATGATGGTGGTTATTATACCGTCAAGGACTGT
GTGACTATTGACGTCCTTCCCCGTACGCCCGGCAATAACGTCTACGTTGGTTTCATGGTTTGGTCTAACTTTACCGCTACTAAATGCCGC
G END H START
GGATTGGTTTCGCTGAATCAGGTTATTAAAGAGATTATTTGTCTCCAGCCACTTAAGT[illegible]TTATGTTTGGTGCTATTGCTGGCG
GTATTGCTTCTGCTCTTGCTGGTGGCGCCATGTCTAAATTGTTTGGAGGCGGTCAAAAAGCCGCCTCCGGTGGCATTCAAGGTGATGTGC
TTGCTACCGATAACAATACTGTAGGCATGGGTGATGCTGGTATTAAATCTGCCATTCAAGGCTCTAATGTTCCTAACCCTGATGAGGCCG
CCCCTAGTTTTGTTTCTGGTGCTATGGCTAAAGCTGGTAAAGGACTTCTTGAAGGTACGTTGCAGGCTGGCACTTCTGCCGTTTCTGATA
AGTTGCTTGATTTGGTTGGACTTGGTGGCAAGTCTGCCGCTGATAAAGGAAAGGATACTCGTGATTATCTTGCTGCTGCATTTCCTGAGC
TTAATGCTTGGGAGCGTGCTGGTGCTGATGCTTCCTCTGCTGGTATGGTTGACGCCGGATTTGAGAATCAAAAAGAGCTTACTAAAATGC
AACTGGACAATCAGAAAGAGATTGCCGAGATGCAAAATGAGACTCAAAAAGAGATTGCTGGCATTCAGTCGGCGACTTCACGCCAGAATA
CGAAAGACCAGGTATATGCACAAAATGAGATGCTTGCTTATCAACAGAAGGAGTCTACTGCTCGCGTTGCGTCTATTATGGAAAACACCA
ATCTTTCCAAGCAACAGCAGGTTTCCGAGATTATGCGCCAAATGCTTACTCAAGCTCAAACGGCTGGTCAGTATTTTACCAATGACCAAA
TCAAAGAAATGACTCGCAAGGTTAGTGCTGAGGTTGACTTAGTTCATCAGCAAACGCAGAATCAGCGGTATGGCTCTTCTCATATTGGCG
CTACTGCAAAGGATATTTCTAATGTCGTCACTGATGCTGCTTCTGGTGTGGTTGATATTTTTCATGGTATTGATAAAGCTGTTGCCGATA
H END
CTTGGAACAATTTCTGGAAAGACGGTAAAGCTGATGGTATTGGCTCTAATTTGTCTAGGAAATAA

FIGURE 9.4 Example of the DNA sequence of a virus.

information and of yet executing the plan. There must be natural laws according to which an organism develops from a given DNA.

We have compared the DNA to a tape recording on which magnetic pulses are stored. What we also have to know is the analog of the cassette player, which converts the signals into a tune. There is one very marked difference: everything suggests that nature converts the DNA signals in an unbelievably ingenious way, as it were prescribes only the theme of the piece of music, but leaves the detailed development to the instrument—i.e., to the growing organism. But this calls into question the claim that the DNA contains closely defined items of information. It depends entirely on the environment in which the DNA (or RNA) "plays back" its theme. To take an extreme example, if we place the DNA or RNA in a pile of sand nothing at all will happen. On the other hand, certain fragments of these substances can already "enslave" certain bacteria to manufacture insulin.

Model Examples of Biological Formation

Before we pursue these questions further, however, we must again turn to experiments which can explain the mechanisms in the development of forms or organs. In biology as in all other branches of science, certain model systems are used whose properties are relatively simple to study. Two very widely known examples are the slime mold and the hydra.

The slime mold normally exists in the form of individual ameboid cells living on a substrate. If the food for the individual cells becomes scarce, these cells suddenly congregate, as on a secret word of command, in a certain point, where they accumulate more and more and differentiate into stalk and spore case (fig. 9.5). The slime mold as a whole is capable of locomotion, writhing on the ground like a snake (fig. 9.6). The first phase, that of accumulation, is immediately highly interesting. How do the individual cells know that they have to assemble and where? As the biologists discovered, the cells can produce and secrete a substance (the so-called cyclic adenosine monophosphate, cAMP). If a second cell comes into contact with cAMP, it can secrete more cAMP. Through the cooperation of this reinforcing effect on the one hand and diffusion on the other, patterns of chemical waves or chemical spirals (fig. 9.7) are created. The individual cells are able to measure the density gradients of the cAMP waves, moving against the

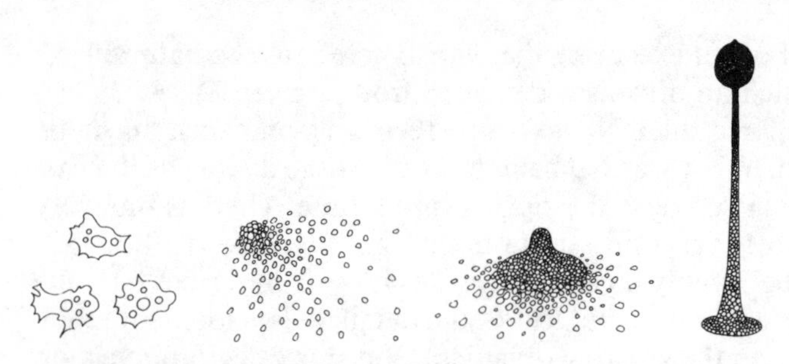

FIGURE 9.5 Schematic representation of the development of a slime mold from individual amoeba to the adult fungus.

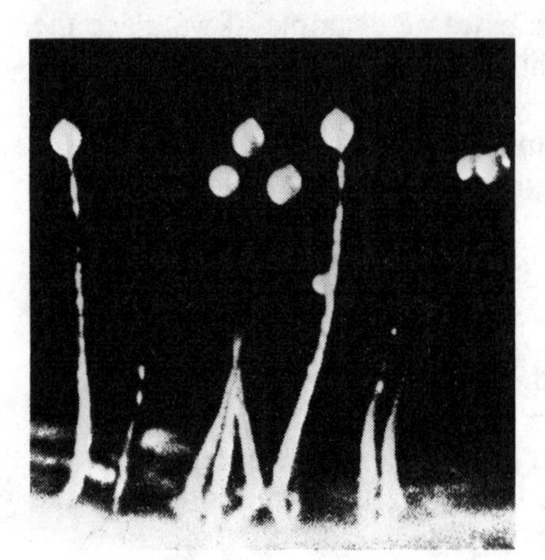

FIGURE 9.6 Slime molds.

direction of the gradient by means of small protrusions that they use as paddles.

This example shows us quite clearly that pattern formations such as spirals or concentric rings can proceed in complete analogy both in chemical reactions in inanimate nature and in animate nature (here in the formation of slime molds). The underlying cause is the fact that the same laws for the order parameters that describe the macro-order also govern the pattern formation.

After the individual identical cells have assembled, a new process starts that can be readily observed but whose various causes have not yet been fully explained. The cells adhere to each other; on one side of

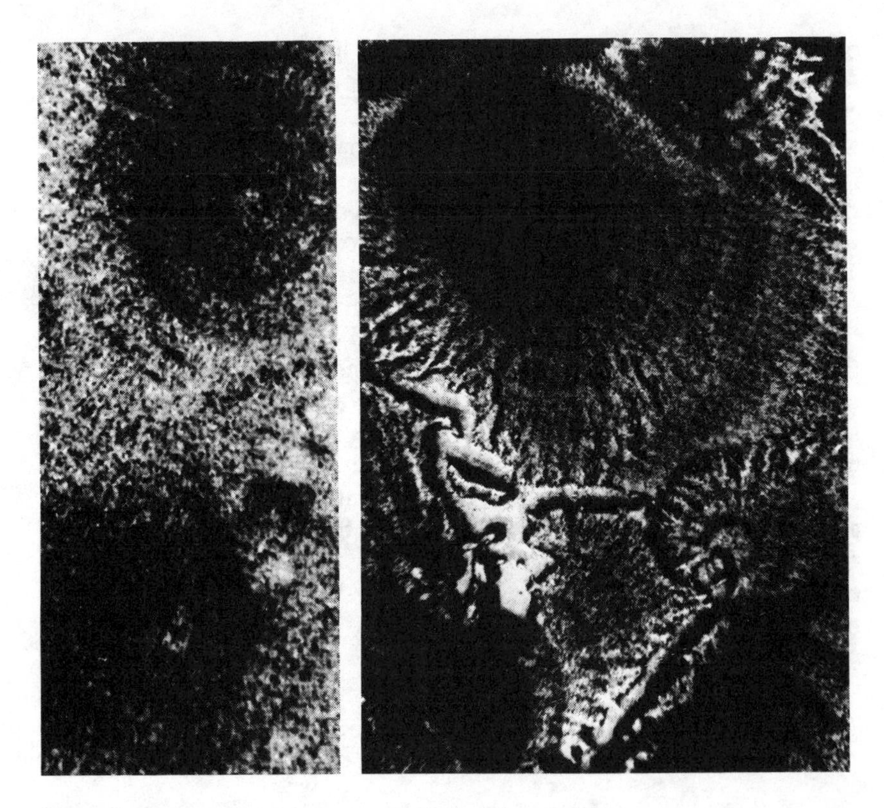

FIGURE 9.7 The spiral waves of cAMP, (a) and (b).

the accumulation they form a stalk, on the other a spore case; i.e., they differentiate. The cAMP appears to play a decisive part in this process of differentiation also, but research into the quesiton is not yet complete. This example nevertheless demonstrates quite clearly that the individual cells communicate with each other through the medium of a chemical substance. Realizing this will immediately help us to an understanding of pattern formation.

A very well-known example is the hydra, a freshwater polyp measuring a few millimeters and consisting of a few hundred thousand cells of about a dozen different types. The hydra has a head and a foot. The question we want to study is how a previously undifferentiated union of cells knows where it should form the head and where the foot. According to the previously discussed idea of a pre-existing plan, we could assume that at the outset each cell has received its instructions about what is eventually to become of it, head or foot.

The following experiment can be conducted with the hydra (fig. 9.8). If the animal is cut through the middle, two new animals form; the one with the head will regenerate its foot, the one with the foot its head. This shows that identical cells can develop into entirely different organs. They must therefore somehow receive their instructions from the union of cells and learn where they are—at the end which is to be the head or at the other, destined to become the foot. In other words the individual cells must be able to receive information about their location in the union of cells. Further experiments shed light on the mechanisms that control this communication.

If a part of the head of a hydra is transplanted into the middle part of another hydra but close to the old one, the formation of the new head regresses. But if the newly implanted head is far enough away from the old one, it will grow into an entirely new head. Obviously the cells must be capable of communicating across fairly long distances: an existing head ensures that no second head will grow in its close vicinity.

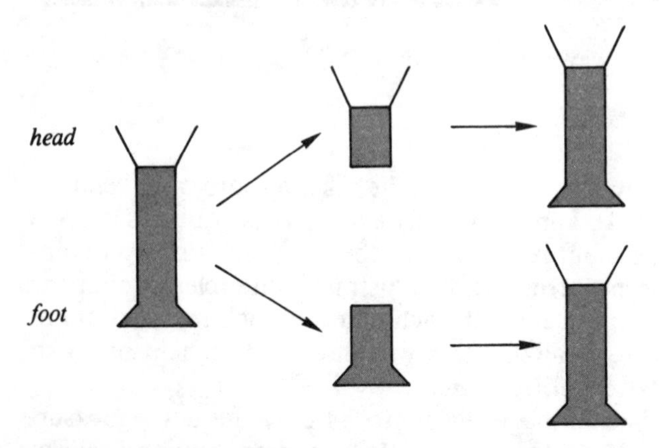

FIGURE 9.8 Diagrammatic representation of the regeneration of a hydra. *Left:* Intact animal with head and foot. *Center:* The two parts of a hydra after separation. *Right:* Regeneration of the foot (*bottom*) and of the head (*top*).

Macropatterns on a Molecular Basis

The example of the slime mold has revealed that long-distance communication between cells can be established by diffusing chemical substances. Indeed, mathematicians (A. M. Turing) suggested models long ago to explain the phenomenon of cell differentiation. Let us look at two cells, originally separate from each other, in which the same chemical processes run their course (fig. 9.9); a type A molecule is produced only to be partly decomposed until an equilibrium concentration is established, obviously the same in both cells. We now allow the substances to commute between the two cells in both directions (fig. 9.10). The exchange of matter may render the state of equilibrium concentration in the two cells unstable. This can again be best demonstrated with a ball on top of a hill—i.e., with a synergetic curve. If the ball rolls to the left, it means that the concentration of the type A molecules increases in the cell on the left; if to the right, then type A molecules increase in the cell to the right. A minor initial variation in the production of the type A molecules decides which of the two cells will have the higher concentration. Whereas in the separate cells the concentration was the same, that is the type A molecules were symmetrically distributed among both cells, this distribution is asymmetrical in the linked cells; i.e., the symmetry is broken. It is this kind of spatial break of symmetry that has gained such importance in the modern theories of pattern formation. A number of research workers developed Turing's basic idea further by designing special models of chemical reactions in very many cells. More strictly, they investigated reactions in a continuum.

Alfred Gierer and Hans Meinhardt designed a detailed mathematical model that can explain, for example, the formation of the head and the foot in a hydra. It deals particularly with the question of how, in an initially undifferentiated elongated union of cells, a head can form at one end and a foot at the other end. Let us imagine an initially undifferentiated union of cells in which two different substances are produced; one of them stimulates the formation of the head and is therefore called stimulant. But we have just seen that the formation of the head can also be inhibited, and we must postulate a second substance that prevents or inhibits the growth of a head, to be called inhibitor. Let us assume that at first the cells of the union uniformly produce stimulants and possibly also inhibitors, that these substances can diffuse through the cell union, and that they react with each other. This

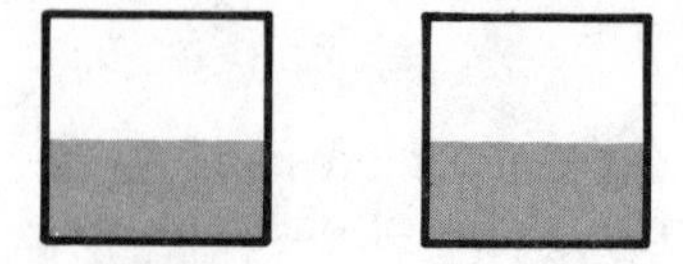

FIGURE 9.9 In separate but identical cells, the same concentration of a chemical substance is produced.

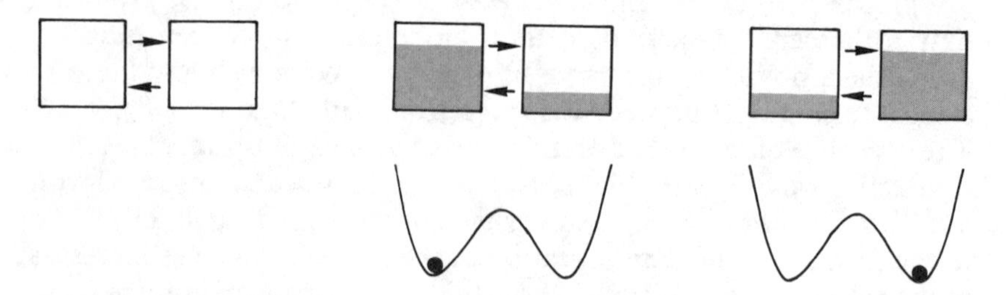

FIGURE 9.10 The cells of Figure 9.9 are moved together, which allows an exchange of material between them. Together with the reactions in the individual cells, this produces an unequal distribution of the concentration of the material. The break of symmetry that has occurred here is symbolized by the two positions of a ball in the double bowl.

leads us back to the investigation, as in the chapter on chemical reactions, of the combined effect of reaction and diffusion processes. We are no longer surprised to find here, too, a chemical pattern at certain critical rates of production, say of the stimulant. The result can, for example, be a concentration gradient, which can perhaps be the simplest imaginable pattern (fig. 9.11). According to current ideas, a high concentration of the stimulant is said to be capable of switching on the genes of the individual cells located there, which effect the differentiation of the cells to produce the head. The sequence of processes thus fits perfectly into the general scheme of synergetics. The ultimately developing chemical pattern is the order parameter, created on the one hand by the cooperation between the chemical substances, but on the other controlling the sequence of the individual processes to ensure the creation of this particular pattern.

As in all the previous examples, the order parameter can be represented in two ways: either by the pattern in space (or in time) immediately accessible to our perception, or by exact calculation. As soon as we describe the processes in the form of the already mentioned reaction-diffusion equations, the methods of synergetics yield the resulting distribution of concentration. Synergetics also shows that entirely different reaction sequences can produce the same spatial pattern (fig. 9.12).

For some time the stimulants and inhibitors postulated here looked purely hypothetical, but they have now been verified by experiments: a stimulant and an inhibitor exists each for the head and for the foot of the hydra.

These stimulants and inhibitors seem to be widespread in nature. For example, they have been demonstrated in the sea anemone, and they even play a decisive part in the development of the mammalian nervous system. A factor for the growth of nerves was found some time ago. Secreted by cells, the substance diffuses through the nervous tissue, attracts nerves that grow from another union of cells, and controls their growth toward, let us say a peripheral region of the body.

The fundamental principle of the stimulant and the inhibitor on the one hand and the regulation of pattern formation with the aid of macrocontrols on the other offers at least a basic explanation of many phenomena, such as the formation of stripes on the zebra, the development of buds on shoots (fig. 9.13) and many others. We are, howev-

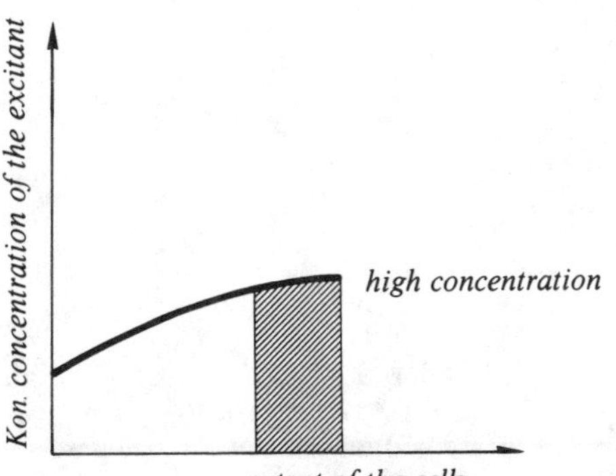

FIGURE 9.11 Example of the distribution of the concentration of biomolecules. *Left:* Low concentration. *Right:* High concentration.

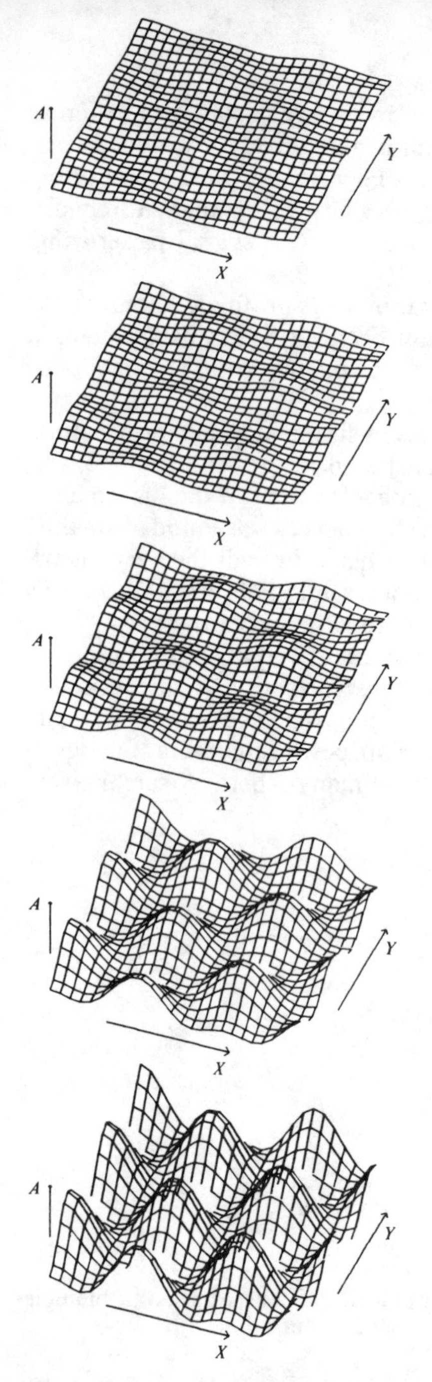

FIGURE 9.12 Concentration of an excitant entered on top of a two-dimensional cell union. The gradual increase in the concentration of the excitant and its concomitant pattern formation is investigated with the aid of a mathematical model.

er, without doubt still at the beginning of a long development before we can understand what is involved in the production of complex organs like the heart or the eye.

These examples reveal clearly that nature proceeds, on the basis of the building plan enshrined in the DNA, in a far more sophisticated manner than was originally thought. Nature lets the individual parts of the growing organism communicate with and balance each other. Presumably it uses this trick for the formation of at least part of the brain, which will be discussed in Chapter 14. Here, too, the parts do not by any means grow according to a fixed plan such as we would use for the construction of an electronic instrument. Processes of self-organization are also involved in the development of nerve tracts between the sense organs and the brain. This is suggested by experiments in which the nerve tract between the eye and the brain of a frog was severed and allowed to reconnect so that, in one instance, part of the brain saw the surroundings upside-down. After a short time, however, we could gather from its behavior—e.g., the way it caught flies—that the frog regained its ability to see correctly. The function of the connections must have changed to ensure the reestablishment of a uniform "correct" transmission scheme.

This raises the fundamental question of how the visual cells of the eye are connected with the corresponding nerve cells in the brain even during growth. A more detailed discussion of these experiments shows

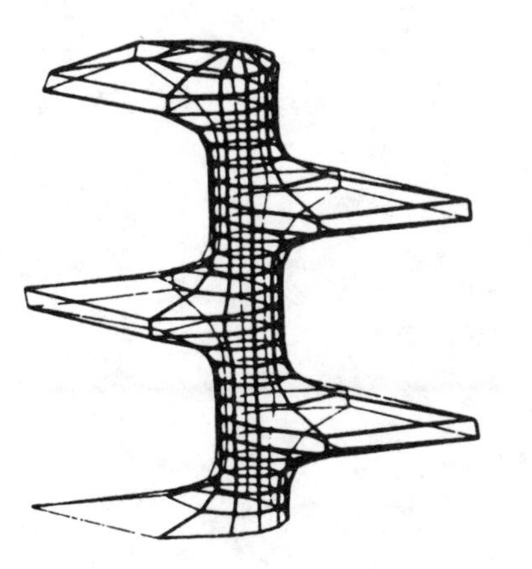

FIGURE 9.13 Example of a mathematical model of bud formation on a stem.

that the connections between the eye and the brain organize themselves. As model computations by Christoph v.d. Malsburg reveal, this is also controlled by the principle of competition; accordingly, small regions that form the image correctly are reinforced and other nerve fibers that produce the wrong image are suppressed. "Correct" and "incorrect" in this context mean that the image of the surroundings in a zone of the eye is formed on a corresponding contiguous zone in the brain. Cooperation and coexistence on the one hand and competition on the other are therefore phenomena that are by no means confined to the macro animal kingdom. The individual organism time and time again develops according to this basic principle as well.

Chapter 10

Conflicts Are Sometimes Inevitable

On many occasions we noted a peculiar feature in our examples from physics. Whenever a new state of order begins, nature leaves the system a choice of several possibilities. For example, when a flow pattern is produced in a liquid heated from below, the rolls are equally likely to revolve clockwise or counterclockwise. As already pointed out, a simple mechanical model provides an easy explanation of this behavior.

If we place a ball in a bowl shaped as illustrated in figure 10.1, the ball will drop from its unstable position into a new one, but the two new positions are completely equivalent. The existing symmetry must be broken by the position that the ball assumes.

The task of determining which position the ball would assume has no definitive solution. Quite obviously it has two entirely equivalent ones. This contradicts our normal feeling that every problem must

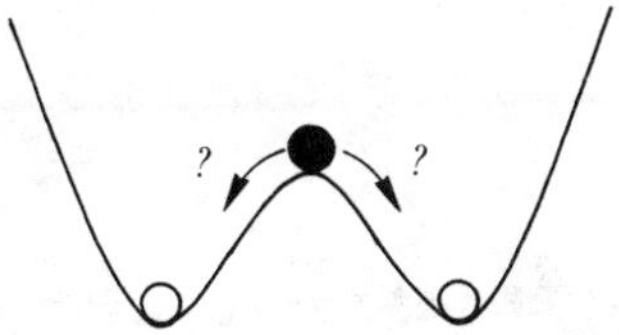

FIGURE 10.1 The familiar model for a problem with two equivalent solutions. Where will the ball roll?

have a clear-cut solution. A little experiment that we can conduct on ourselves shows readily that this kind of problem is not confined to mechanics or to simple natural processes.

There is no doubt that our brain is the most complex system ever created by nature. Here, too, we have the phenomenon of the symmetry break, for instance in perception. When we look at figure 10.2, it

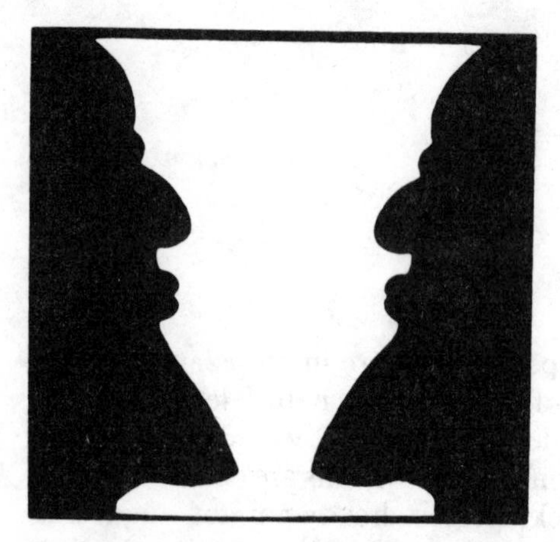

FIGURE 10.2 Vase or face?

FIGURE 10.3 Angel or devil?

will at first perhaps appear meaningless. But if asked to regard the white center portion as the foreground, we shall at once perceive a vase. Yet if told to regard the two black portions as foreground, we shall at once perceive two faces. The perception contents of the picture are ambiguous, and the perceptions of "vase" and of "faces" rank equally. A large number of pictures by the now very famous artist M. C. Escher are based on this broken symmetry. An example of his many works is illustrated in figure 10.3, in which we recognize alternatively angels and devils. These illustrations show that we need additional information to break existing symmetry in perception, such as the information "regard the white portion as foreground." But even without additional external information we are usually able to break the symmetry. We do this by a process in our brain that, analogously to phenomena in physics, say in liquids, we can call a fluctuation. Suddenly a perceived image builds up, the perception occurs like a flash.

Test Your Own State of Mind

Symmetries in our brain can be broken from the outset with a psychological imprint; i.e., we harbor a kind of unconscious prejudice. A number of psychological tests are based on this fact.

The reader can test his or her own reactions with the faces illustrated in figure 10.4. Are they sad or cheerful? Is there a relationship between the two women? If so, what is its nature? Look at the picture carefully before you make your statement. The pictures have in fact been drawn so that their expressions are neutral, neither sad nor cheerful. But with this test, the demand for a decision means the inherent symmetry of this picture must be broken. It is possible only on the basis of additional information, which is provided by the subject of the test according to the state of mind he or she happens to be in; this is projected onto the persons in the picture. These tests allow the psychologist to draw conclusions about the unconscious state of mind of the person tested.

We can describe this process in other words. We have read into the pictures that quality for which we had already been inwardly prepared. We often hear or otherwise deduce what we happen to expect. In a certain sense this is even necessary, because perception implies that we constantly link fresh experiences with those in the past already stored as "experience."

FIGURE 10.4 The viewer projects his or her state of mind onto expressionless faces or neutral sentences. Picture from the Thematic Apperception Test developed by Henry Murray and coworkers at the Institute of Clinical Psychology, Harvard University, in 1935. This projective test, similar to the Rorschach Test, is designed to discover subconscious conflicts and thoughts.

I remember another test from my school days; it consisted of an impossible task. A student was asked to copy a text; it was absolutely essential that it should be completed and delivered quickly, and in copperplate handwriting. But the task was set up so that the text, if carefully written, could not be completed and delivered in time, or else it could be complete but badly written. A typical conflict situation, one only too gladly evaded. But it is constructed so that to solve the task the test-taker is forced to make use of the symmetry break incorporated in it. The psychologist hopes that the decision, one way or the other, will give him information about the character of the person tested, whether he or she is slovenly or neat. Obviously the whole test become farcical if the person knows the psychologist's intention from the start and fools him.

Tests of this kind are performed everywhere, for instance in the United States Army. The following story has long circulated among physicists about a very prominent American colleague. He was about to be called up for army service, which involved taking a psychological test. To gauge his sincerity the psychologist asked the recruit to show him his hands. If he showed them palms up, the psychologist would conclude that the recruit's character was sincere. If he showed them backs up, the psychologist would claim to have taken his measure as a secretive character. But at the psychologist's instruction, the physicist reacted by showing one hand palm up, the other palm down. This gave the psychologist a minor shock; he quickly suggested, "For heaven's sake, man, turn your hands over," whereupon the physicist turned both hands again with one palm facing up, the other down. Result: physicist duly rejected by the army. Perhaps it is the psychologist who will now seek advice.

Life Is Full of Conflicts

We have so far discussed only artificially produced conflicts; but life abounds with the natural kind, as a few examples will show. A young man would like to study but vacillates between two entirely different subjects. Each of the two courses offers him both advantages and disadvantages.

Another example concerns a young girl. As luck would have it, she meets two very nice men in quick succession, both of whom want to marry her. She feels attracted by both and does not have the heart to refuse either. She is torn between them. Finally, a single sentence uttered by one tilts the balance, and spontaneously the girl accepts him. In synergetic terms a fluctuation—a single sentence—has tipped the scales.

But the following situation is found more frequently in the field of psychology. A widower finds his loneliness hard to bear and wants to marry again. His eagerness is rewarded: within a short space of time he meets two ladies who would not be averse to marrying him. But now he begins to wonder: which should he marry? In many discussions with friends and acquaintances he tries to balance the pros and cons of the two ladies. But this process leads him deeper and deeper into a conflict. He finds that he is unable to make a decision. Advantages and disadvantages, which naturally also influence his affection for each, balance each other. This is a typical example of psychological conflict that prevents a decision from being made; he vacillates until both ladies lose interest in him.

We can illustrate the psychological reaction quite clearly with our mechanical model of a ball in a bowl, figure 10.5. But let the ball be made of steel and the bowl of a relatively soft material. The longer the ball remains "undecided" in the center, the deeper it will sink into the material of the bowl until it is captive in its self-created depression, unable ever to leave it. In the psychological case of course it may be that the indecision is the result of a subconscious refusal to marry and the conflict is merely a stratagem to reach, like the ball, an irreversible position.

In times of severe threat such indecision may become fatal; in his book *Erziehung zum Ueberleben* (Education for Survival), Bruno Bettelheim gives trenchant examples: persecuted individuals continuously vacillating between hiding from the regime and escaping from its clutches by a daring flight.

These examples are typical of many conflicts. At first, two solutions appear equivalent. We then start to speculate about them by looking for higher-order aids to decisionmaking designed (in the language of synergetics) to break the symmetry. We are now faced with the important realization that springs from the essence of synergetics: in a number of cases no higher-order aids to decisionmaking exist at all. In fact,

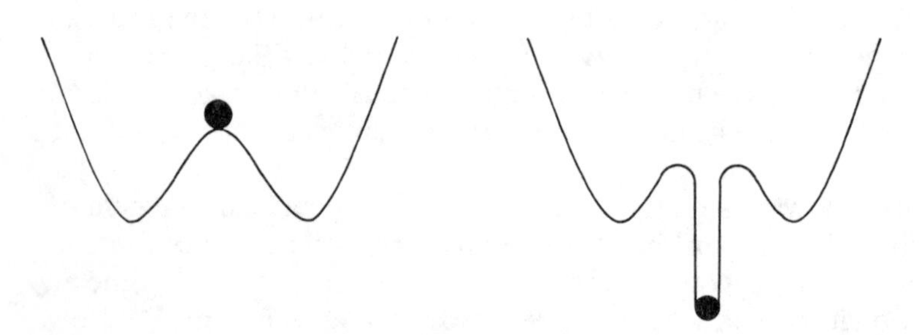

FIGURE 10.5 Demonstration of the effect of too much pondering. The ball is digging a hole from which it can no longer escape.

even after the most prolonged speculation we are unable to resolve the conflict once and for all. Furthermore, life offers us all too little time to arrive at a decision. This allows only one conclusion: we must accept that there are problems that basically admit of two (or even more) equivalent solutions. We will often find it impossible to decide whether they are in fact basically equivalent, and, after a period of speculation during which we tried to balance the various aspects, we must acknowledge that we are facing a genuine conflict and that each of the solutions is indeed equivalent. In such a case, however, the choices are also equivalent. A further consequence is that we should not regret our choice; we should remember that after all our speculations we have chosen one of the equivalent solutions, and not forget that the other solution would also have had its disadvantages.

Conflict Displacement in the Social Field

Especially in the social field, conflicts exist that offer two equivalent solutions or expedients in which action *by the community* resolves the conflict for individuals not by removing it but by merely displacing it. Here are a few examples that, although relatively innocuous in themselves, provide analogies with quite explosive problems: when a child is born, as a matter of course it must have a surname. In many countries it is the custom or even the law that it should have its father's name. But it could equally well have its mother's name. The absence of legal regulation would create a conflict situation for all parents: "Which of our names is the child to have?" There is no doubt that this is a situation of potential conflict for all marriage partners; in the absence of a legal regulation, they would have to reach an agreement between themselves.

The same applies to marriage. Are the partners to take the husband's or the wife's name? Some couples agree on a double name: Jones-Brown. It is easy to see that after ten generations there would be more than a thousand individual names. Nonsense indeed, which reduces this compromise to meaningless absurdity.

We are in fact left with only the two possibilities—adopting the name of either husband and father or wife and mother respectively. Earlier societies obviously recognized this problem, and the attitude of the community has broken the symmetry by either tradition or legislation. The law assumes the role of order parameter by enslaving the marriage partners in naming the child. But on the other hand, at least in democracies, laws are promulgated by the people's representatives; the order parameter is determined by the individual citizens. Once more we recognize the typical interrelation between order parameter and individuals that we encounter in synergetics. The other possibility is the absence of the order parameter. Here the law leaves the naming open and passes the buck to the family. This forces us to realize that greater individual freedom means greater individual conflict possibilities.

Another example of such conflicts is provided by parental rights. Do the courts favor the father or the mother in the child's upbringing? This question is particularly obvious in divorce cases, where as a rule the judge awards the custody of small children to the mother. Here again symmetry is collectively broken. In fact, the father could also be awarded custody. In that case, the symmetry must be broken by the judge. In the absence of the necessary laws the arbitrary breaking of the symmetry has merely been transferred to another decisionmaking body. The following example may shed yet more light on the complex of problems: the question of whether partners are better off in a proper marriage or living in a marriagelike partnership. As in typical conflicts, the advantages and disadvantages of one solution are balanced against those of the other. Among the large number of such polarities of advantages and disadvantages are, in marriage, the constraint on the one hand and the possibility of being cared for on the other. In a marriagelike partnership there is freedom, but it automatically implies that there is no obligation on the part of one partner to look after the other. We must be fully aware that it is impossible to enjoy both advantages at the same time. The breakup of a marriagelike partnership sometimes leads to considerable financial arguments, as when resources had been pooled to buy a house or apartment, and arbitration is left to the authorities. An attempt is made to displace the conflict to the collective—i.e., the authorities, who often have made no provisions for such arbitration because the very existence of the institution of marriage was meant to deal with such questions.

All these examples, which could be added to at will, illustrate that in a political community conflicts are constantly shifted from the individual to the collective or from the collective to the individual. This interrelation between the individual and the collective results in the individual being relieved of personal decisions that may produce conflicts by collective measures such as the law. Conversely, a greater freedom of decisionmaking for the individual means an increase in the possibilities of conflict.

Such collective measures are not restricted to marital affairs. They may affect whole communities or cities without any chance of legislative measures becoming effective.

If you have lived in several different localities for some time you will find that the climate of human relations varies from place to place. In some cities or towns the people are very friendly, while the inhabitants of others are downright rude to each other and to strangers. Here the choice between the personal attitude of "friendly" and "rude" is obviously a break of symmetry made collectively. Once a general mood has become established in a certain locality, a newcomer will be unable to prevail against it, and in the course of the years his behavior will often differ little from that of the natives. If he tries to be friendly in an unfriendly place, he will soon become frustrated and probably unfriendly himself. Conversely, when an unfriendly person moves to a friendly place he will as likely as not be infected by the general friendliness of the inhabitants. We observe the same phenomenon not only in whole localities but also in offices and expressed by management, sometimes creating an even more powerful factor. Local atmospheres, against which a newcomer will only very rarely be able to prevail, may vary widely here, too.

Chapter 11

Chaos, Coincidence, and the Mechanistic World Picture

Predestination or Accident?

Hardly a philosopher, let alone a natural scientist, would argue that the knowledge gained through physics and the natural sciences in general has not deeply affected the picture we have formed of our world. Our whole thought has been profoundly influenced by the scientific revolution, which has shaken physics to its very foundations. Only the laws of physics and their millionfold confirmation have reinforced our conviction that the natural processes are governed by immutable laws. The flourishing of mechanics during the nineteenth century made an important contribution to this belief.

Mechanics investigates how individual bodies move on the basis of the forces acting between them. It was Sir Isaac Newton's fundamental realization that the fall of the apple from the tree is governed by the same laws as are the earth's and other planets' orbits around the sun. For instance, Newton's laws are the basis of all rocket technology; the basis, then, of man's conquest of space. As we can observe on television, the rockets to the moon follow precisely precalculated paths. But this adherence to a predetermined and therefore predicted path appears sinister, even oppressive to us; if the sequence of the most various events is predestined, we can be no more than part of an immense gear mechanism, by definition willy-nilly surrendered to it.

Accident itself has no chance; everything is predestined. The profound philosophical and religious implications of such a picture of the world have often been discussed; they are readily imagined. But during the twenties the quantum theory represented an unexpected turning point. The accident came back into its own.

Let us briefly recall what happens in a lamp or in a laser. If we excite an electron in an atom—i.e., it receives more energy than it normally has—it tends to get rid of the energy by emitting it as a light wave. As the quantum theory shows, it is fundamentally impossible to predict exactly when the electron will radiate its energy. This is much as in a game of dice, when we cannot predict the number of dots we will throw.

All that we know today of the events in the microcosm, in other words, the invisible world of atoms, says that those events are governed by accidents, by random occurrences. All attempts to restore the mechanistic picture of the world to its former position have failed; they have led us into direct conflict with experimental evidence. The accident as something unforeseeable grossly contradicts the concept of events progressing along once-and-for-all predetermined lines.

Predestination and Accident!

It was a great surprise to many scientists when it became increasingly clear in the last few years that there are events in many aspects of nature that occupy a kind of intermediate position. For example, kinetic events on the one hand obey laws as immutable as mechanics or are themselves laws of mechanics. On the other hand, there is something random and unpredictable inherent in them. This entirely new group of phenomena, which is only now beginning to enter the general consciousness of the scientific world, is called chaos.

We are familiar with this word in our everyday language. The chaos on the roads, an inextricable tangle of vehicles, a hopeless confusion, comes most readily to mind. This example reveals the characteristic traits of the word "chaos" as now used in the scientific sense. Every single vehicle has, after all, reached its place in the traffic jam according to the strict laws of mechanics. The traffic jam—the chaos—nevertheless appears to the observer as something totally confused, with the positions of the individual vehicles seeming to be distributed as if at random. A truck next to a blue sedan, a red sedan crossing its path, a motorcycle behind it, and so on. The only different from the chaos of natural events is the fact that in the latter—figuratively—the vehicles are all in motion and continuously change the positions into which they have been wedged. The chaos, initially regarded as an absurd, isolated case, now appears to us as a typical behavior pattern of many systems we investigate in synergetics. Let us recall a few examples we have met before.

In the motion of liquids, widely different motion patterns are produced depending on how much we heat a horizontal layer of a liquid from below. After a few stages, at which uniform motion patterns such as rolls or honeycombs have formed, the liquid assumes an entirely irregular motion. It becomes, as the expert says, turbulent. We are now justifed in assuming that this confused, completely random motion is subject to the laws of chaotic motion.

We observe similar phenomena when smokers blow smoke rings into the air. First the rings become deformed and finally a quite irregular motion of the smoke begins. It has become turbulent. Macropatterns are produced by certain chemical reactions, either in space or in time, such as the periodic change from blue to red, etc., in the Belusov-Shabotinsky reaction.

Even earlier, chemists had observed the completely irregular timing of the change from red to blue. But they thought they had not prepared the mixture properly and refrained from publishing their findings. Now, after the phenomenon of chaos has become generally acknowledged, the chemists are vying with each other to find and to publish more and more new results of irregular time sequences and spatial patterns of such reactions. There are even predictions that laser light could be turbulent—quite irregularly emitted wave trains, yet of a character different from that of the light of ordinary lamps. An entirely new kind of light is waiting to be discovered.

In biology, too, the idea of chaos is beoming accepted; it has suddenly thrown light on previously quite incomprehensible phenomena, such as the totally irregular fluctuation of insect populations from one year to the next. Models already exist according to which these fluctuations can be mathematically interpreted.

All these phenomena, which to most of us seem entirely new, are aptly summed up for the scientists in the words of Ecclesiastes: "There is no new thing under the sun."

As far back as the turn of the century the French mathematician

Jules Henri Poincaré (1854–1912) had discovered the possibility of chaotic motion in completely different areas—i.e., in celestial mechanics. If we investigate the model of a solar system that has two suns but only one planet, this planet may execute the most incredibly complicated motions, like a football propelled with random kicks. Here the dilemma of science becomes evident. This motion of the planet, after all, conforms to the strict laws of mechanics, yet it nevertheless appears purely random.

This example shows that even a perfectly simple mechanical system is capable of executing extremely complex motions. In the past, we took it entirely for granted that—according to Newtonian mechanics—planets in our solar system move around the sun in elliptical orbits, permanently and immutably. But today, this constancy of planetary orbits appears puzzling in light of the development of modern mechanics. The most eminent scholars have considered this problem in order to find the answer to the King of Sweden's prize question of the last century: "Is our solar system stable or is it possible that some planets plunge into the sun, for example, and others are ejected from it?" All these events are compatible with the energy and momentum theory of mechanics.

The answer mathematicians offer to this question today is so subtle and linked to such fine details of the planets' revolutions that it is sometimes hard to believe that it is the final one. Nonetheless, if their theory is correct it should be possible to explain the structure of Saturn's rings (fig. 11.1). The rings, thought to be made up of small lumps of ice, have so far appeared through the telescope to be individual concentric rings. This poses the question of why there are gaps between them.

Why did the lumps of ice vacate those areas to form the gaps? Here is the explanation given by mathematicians engaged in the study of celestial motions: the effect of Saturn's moons forces the lumps of ice in these areas into chaotic tracks; the lumps must therefore leave the areas and by so doing create the gaps. Whether this is the last word remains to be seen. Closeup photographs taken by American space probes in fact reveal even much finer structures. The rings of Saturn appear to be grooved like a phonograph record, and something like spokes seem to exist in the gaps that hitherto were thought to be empty. Unsolved riddles.

An unequivocal answer to the question why chaotic, random

FIGURE 11.1 The rings of Saturn.

motions can come about is possible only within the framework of mathematics, which has not yet progressed beyond the initial stage of comprehending the chaos.

We can nevertheless very readily visualize how chance can furtively enter into strictly predetermined motions.

Fruit-Sorting Machines: Planned Chaos

Let us imagine a sharp edge, such as a razor blade, which we set up vertically and allow steel balls to drop on (fig. 11.2). Whether the razor blade deflects a ball toward one side or the other depends on the most minute fractions of a millimeter of its impact on the steel ball. Very little to the left below the center and the ball will be deflected to the left, and conversely to the right. The entire process is obviously strictly predetermined; nevertheless there is something random about it. The reason: we are unable to predetermine or to measure the initial position of the ball with perfect accuracy. But a very minor displacement from the ball's original position will result in a totally different

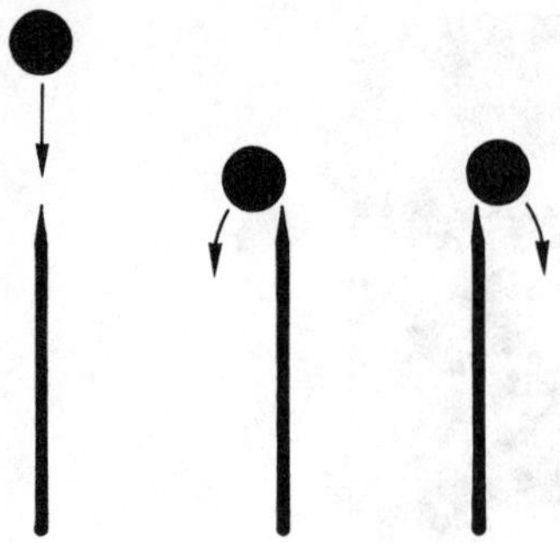

FIGURE 11.2 A steel ball dropping on a razor blade.

path. Exactly the same applies to a game of dice. A die generally hits the surface of the table with one of its edges; what happens next depends every bit as much on preconditions as with the balls dropping on the edge of a razor blade.

We can see now that the difference between random and strictly predetermined events begins to blur, although both limiting cases can be strictly defined in philosophical terms, and only these two cases should "in reality" exist. The decisive point is that minor inaccuracies of the original position have a macro-effect on the subsequent course of the event.

Practical people, inventors, amateur mechanics are sometimes cleverer than the most erudite scientists. A whole industry of fruit-sorting machine manufacturers has long been living on the principle that even strictly predetermined mechanical movements can strikingly simulate chance. Such machines, for example, make balls drop on edges. In each action the gambler cannot foresee the path of the ball. The outcome of the gamble is indeed a matter of luck. Nevertheless all the steps happen in an unequivocally predetermined manner. Figure 11.3 is an example of a well-known mechanical fruit sorter.

The North Was Not Always North

Science fiction novels sometimes describe what happens to people who are transported into the future or into the past. Let us assume a science

FIGURE 11.3 Diagram of a mechanical fruit sorter. Where will the ball end up?

fiction author has put a person into a time machine, furnished him with only the barest essentials, and deposited him in a period 100,000 years ago. He must now orient himself with the aid of a compass. Because he feels cold he moves south by the compass. But the longer he travels the colder it gets. It finally dawns on him that he is traveling north instead of south; his compass points in the wrong direction. Since the compass indicates the direction of the earth's magnetic field, we must conclude that the field has changed places.

Obviously we cannot send anyone into the past in a time machine. But nature does it for us in a different way. Geological formations have been found in Greenland that are magnetic. The tiny magnets in the various rock strata were aligned by the prevailing magnetic field of the earth and, as it were, frozen in position; they have maintained their alignment. The stratification also provides a clue to the age of the individual strata. Since magnetization changes from stratum to stratum, it tells geologists that in the course of millions of years the direction of the earth's magnetic field changed several times but at intervals that appear totally irregular. The earth's magnetic field must therefore have reversed its polarity chaotically, as the more recent theories confirm.

Chaos in Synergetics: A Contradiction In Terms?

Having read the last paragraph, some of you may ask what these chaotic events have to do with synergetics. Synergetics, after all, is the theory of cooperation between many parts of a system. When a planet moves in a system with two suns, however, there are only three bodies involved. At the beginning of this book it may have seemed in addition that the cooperation of many individual systems *always* produces ordered structures or reactions. We must therefore deal with these questions in some greater detail, all the more so as this will enable us to draw conclusions about such other fields as economics. To clarify these problems, however, we must adopt a slightly more abstract approach; less than thoroughly interested readers may want to skip the rest of this chapter.

The connection with synergetics will become evident when we remember the concept of the order parameter. A number of examples has shown us that a synergetic system can sometimes be governed not by one but by several order parameters; three order parameters cooperate in the buildup of honeycomb structures in liquids, for instance. They are represented by waves that include an equilateral triangle. In other cases, for example in evolution, various order parameters *compete* with each other. The macroproperties of synergetic systems are thus often described either by the cooperation or by the competition between order parameters.

If the problems of synergetics are formulated mathematically, the same equations are time and time again met with for the order parameters, even when the systems are entirely different. It has been found that certain equations for order parameters can accommodate chaotic processes as well. Once again, in the example of a liquid heated from below: in a phase of chaotic motion three order parameters enter into an interrelation, thereby tossing the system to and fro between its various states of motion.

A more detailed investigation we conducted describes this interrelation between the order parameters as follows: for a certain time one order parameter dominates, enslaving the two others—i.e., prescribing their motions. After a short while, however, it loses its dominant position, yielding it to another order parameter, and the sequence is repeated. This change of domination is totally irregular, that is, chaotic.

The motion of celestial bodies is also covered by this group of equations; here the coordinates of the centers of gravity are the order parameters.

We know now that we must expect chaotic motion in very many interrelations between order parameters and accept as chaos many cases that were explained as measuring errors and/or indignantly rejected on the basis of existing theoretical considerations. Examples: events in the economy, or the effects of administrative interference on largely self-organizing events such as the distribution and development of research and teaching in universities.

Is the Weather Predictable, or Does the Weatherman Always Leave Himself a Loophole?

We may watch television on a Saturday night and plan our Sunday according to the forecast of sunny weather for the following day, only to be bitterly disappointed when it rains cats and dogs. Not only the meteorologists but also physicists and mathematicians have long been concerned with the improvement of weather forecasts. The Hungarian-born John von Neumann (1903–1957), a mathematical genius who later lived in America, invented the basic principles of the modern electronic computers; the first was built in the United States in close cooperation with Neumann during the forties. The mathematician immediately recognized the great technical possibilities of the computer, especially that it was capable of processing an enormous amount of data. He therefore suggested the worldwide establishment of a dense network of meteorological observation posts that would collect data on atmospheric pressure, temperature, wind speed, humidity, and the like, and feed the data into a central weather computer. Because air behaves little different from a liquid, it should be possible on the basis of the fundamental equations of liquid motion to make exact forecasts of the motion of air masses, of their humidity, and therefore of the weather. We have already pointed to the similarity of air and liquid motion with the analogy between streets of clouds and liquid rolls.

Although the network of observation posts has become progressively denser, weather forecasts have hardly improved.

In the sixties the American meteorologist E. N. Lorenz subjected the basic equations of liquid motion to closer examination. Computer calculations led to his discovery that these equations also predict forms of motion that, as we say today, proceed totally chaotically. But what is chaos? To recall the essence of the preceding chapter: a process is chaotic whenever the motion takes a totally different course as soon as the initial values (such as the initial speeds of the masses of air) are changed by a small amount. But we can never measure the movements of air with complete accuracy; even minor measuring errors may produce wildly wrong forecasts within days if not hours.

It seems that the weatherman leaves himself a loophole to surprise us time and time again.

Can Plasmas Be Tamed? Chaos in Nuclear Fusion as Well?

The Greeks counted four aggregate states—earth, water, air, and fire. Three of them are very familiar to us. In modern usage they are called solid, liquid, and gas. But the physicists have in fact discovered a different fourth state, that of plasma.

As we have learned, the various states differ in the microrange only in the mutual arrangement of the molecules they consist of, which in the gaseous state fly freely past each other, colliding only occasionally. If we heat a gas more and more, the movement of the molecules becomes more and more violent and they are torn apart into their original components, the atoms. As we know, every atom consists of the nucleus, which has a positive electric charge, and of a number of negatively charged electrons, which orbit the nucleus. At high temperatures, say a few million degrees, the electrons too assume such violent movement that they become detached from their nucleus with its positive charge. The physicist calls a gas in which the electrons have become detached from their atomic nuclei a plasma. In nature this state is nothing new. Our sun, for instance, is a plasma because of the prevailing temperature of a few hundred million degrees in its interior.

At such high temperatures the individual atomic nuclei collide with each other with such enormous violence that new atomic nuclei can be produced from two smaller ones.

As long ago as the thirties, Hans Albrecht Bethe and Carl-Friedrich von Weizsaecker worked out a scheme according to which atomic nuclei react with each other, with the final effect of producing a new nucleus from four nuclei of hydrogen, that of the helium atom. Where in chemistry the combination of atoms into molecules liberates energy that appears in the form of thermal motion in physics, a fusion can take place between atomic nuclei that liberates enormous amounts of energy. These reactions generate the energy that the sun wastes, one is almost tempted to say, because it radiates the energy into space and only a very minute fraction of it reaches the earth. But this tiny fraction is enough to provide the energy for all the life processes we are concerned with throughout this book.

Because the earth's energy sources—oil, coal, and even the available nuclear energy—unfortunately will be exhausted within the all-too-foreseeable future, we are forced to look for new sources. What would be more obvious than to attempt to copy in the laboratory the reactions in the sun, thereby setting up a minisun as an energy source on earth? The idea of producing plasma for the purpose of nuclear fusion clearly suggests itself.

It is not all that difficult to produce a plasma; in welding arcs, for instance, plasma is produced in the air gap between the electrodes by the powerful electric current. With a number of tricks we can generate high temperatures. But unfortunately the practical realization of nuclear fusion is beset with difficulties. Even at very high temperatures the individual atomic nuclei meet extremely rarely. They must travel many kilometers before they can find a partner with which they can fuse. The plasma would therefore have to have enormous dimensions in the order of dozens of kilometers for fusion to take place. In addition, of course, the particles of a plasma fly rapidly apart. The various constitutents of the plasma, the electrons and the atomic nuclei, would instantly penetrate the walls of the vessel confining them at the enormous speeds they attain at these high temperatures. But physicists have hit upon an idea of preventing this flying apart and at the same time of inducing continual collision of the particles: if the plasma is surrounded by huge magnets, the physicists know the charged particles will be constantly deflected in the magnetic field and forced into a circular orbit. This confines them within a comparatively small space (which may still have a diameter of many meters) and also gives them a constant opportunity of finding partners. The most promising type of plasma machine (fig. 11.4) is called Tokamak, a Russian word: *toka* = flux, *mak* = abbreviation for maximum. It is, then, a machine designed to generate a maximum flux of particles.

This takes us to our central point. Plasmas are an eldorado for research workers looking for instabilities. As mentioned time and time again in this book, instability changes macromotions. The plasma physicists have already discovered far more than a hundred different types of instability—instabilities in which waves suddenly spread in the plasma, instabilities in which entirely new flow patterns occur. We calculated one of them ourselves. Because it is so beautiful, the result of the calculation is shown and explained in figure 11.5. In another example of instability, the plasma flow collapses completely after a short time. The various new waves, flow patterns, etc., are so different that plasma physicists sometimes try to associate them with vital processes. Those plasma physicists who want to put nuclear fusion into practice are, however, less than pleased with most instabilities. If a process continuously jumps from one instability to the next or oscillations can progressively build each other up, it will be impossible for the plasma to be regularly conducted in, say, a ring. A phenomenon is produced—i.e., the aforementioned chaos that has only now entered plasma physics' body of thought. We have given many examples in this book of the possibility of totally irregular motions being set up. It seems that these must

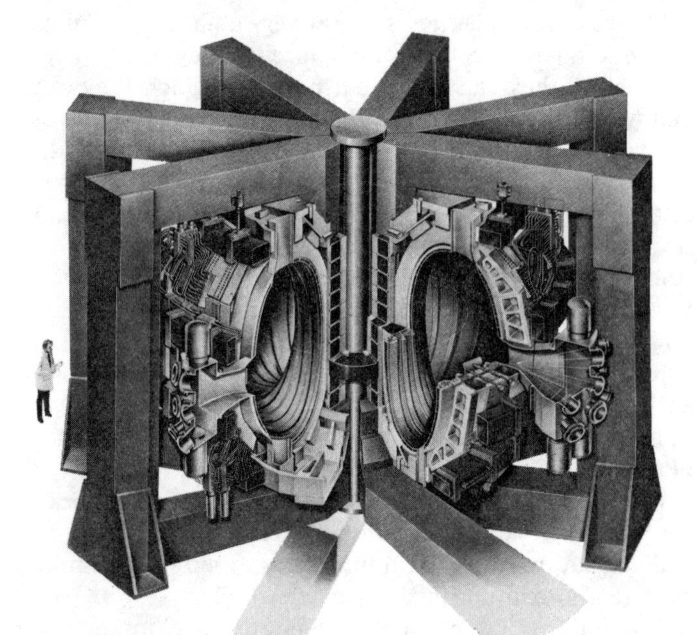

FIGURE 11.4 The gigantic Tokamak machine (its size is indicated by the figure on the left).

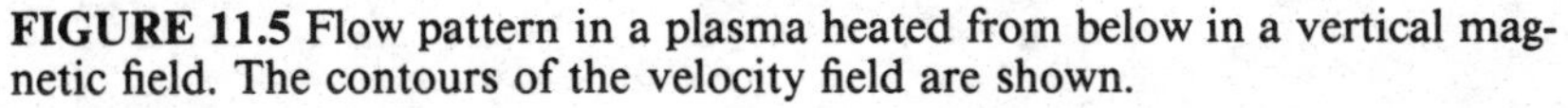

FIGURE 11.5 Flow pattern in a plasma heated from below in a vertical magnetic field. The contours of the velocity field are shown.

also be expected in plasmas. This requires us to understand the nature of chaotic motions so that we can master the chaos technologically. I have no doubt that we shall be successful one day. But a great deal of research still remains to be done; the various disciplines will be able to learn much from each other in the sense of synergetics, because chaos (in the scientific context) represents well-defined phenomena.

Chapter 12

Synergetic Effects in the Economy

In previous chapters we discussed events in the natural sciences, events we were able to define even mathematically. In this and in the following chapter we shall turn to problems concerning human relations. This at once raises the question of whether or not man is a creature so complex that any attempt at establishing a theory that permits prognostication is a priori doomed to failure. In economics, however, we usually deal with the behavior of whole groups, not of an individual. The thesis of synergetics we want to justify and explain specifically claims that with whole groups it is possible to make forecasts.

The question, then, is whether we can describe the behavior of *whole groups,* in either the economic or the social field, on the basis of general laws. By their very existence economics and sociology show that they indeed attempt the scientific exploration of these laws. Here we must realize from the very outset that we shall meet several thought systems because of the difficulties presented by the problems. Two of these systems are particularly prominent. One attempts to understand the whole complex in terms of the behavior of the individual, with the psychological component playing an enormous part. The other deals with events in economics and sociology from its own point of view, and it first must be made clear in our own minds what a "system" means in this context.

To clarify the stance taken by synergetics here, let us look at some concrete examples from economics.

Commercial life as a whole rests on the usually tacit assumption of the existence of laws. Every businessman must make plans, sometimes even long-term ones, and anticipate the behavior of his customers. Let us take an extreme case. A shopkeeper who sells trousseaus will know only in the rarest of cases when a *particular* couple will get married. Nevertheless he is able to predict demand quite accurately, because what is decisive is the action not of an individual couple but of very many couples. It becomes simply a matter of experience to predict that a certain average number of couples will get married per year. The shopkeeper can plan in even greater detail to allow for seasonal fluctuations, which he also knows from experience.

The situation is the same with a bank, which must keep money ready for its customers. It, too, does not know when a particular customer will call and how much money he or she will want to borrow or withdraw. It is nevertheless in a position to keep sufficient money available. Here, however, it is necessary to ensure that the bank does not keep too much ready cash, because this obviously creates a loss of longer-term investment.

These two examples illustrate clearly what we know from experience: that there are laws that govern the behavior of a large number of people. From a synergetic point of view it is important here to distinguish between usual and unusual behavior. We have usual behavior when people act independently of each other, that is to say when they do not previously agree, for instance, that next week they will all buy their bread in a particular baker's shop. The laws of large numbers that apply to independent action were established by the brilliant mathematician Carl Friedrich Gauss (1777–1855) during the last century. They predict not only how many articles should be stocked, but also the fluctuations in sales that must be expected.

The situation is totally different regarding collective behavior. This is precisely what research in synergetics is concerned with. By our definition, collective behavior exists when people act as if they had prearranged the behavior with each other. Obviously it is unnecessary for every individual to talk to every other one or even merely to listen to others.

Here, as in the laser, in liquids, and in all the other examples discussed in preceding chapters, situations exist in which an individual is absolutely compelled to act according to a certain new state of order. Particularly drastic examples are economic catastrophes, such as the panic-selling of shares during a market collapse (which merely intensifies the sudden fall of prices), the flight into gold during inflation, and so on. We shall presently look at less dramatic but perhaps even more important and more typical cases in some detail. With the examples from the natural sciences in mind, we shall find it easier to identify the salient points here. We have learned that a change in external conditions will render a certain state of a system unstable and can replace it with a new and sometimes totally different state. The individual components of the system, for example the liquid, will be drawn into the new state; they will be enslaved.

Synergetics has succeeded in generalizing the law of large numbers in a surprising manner, because it enables us to establish laws even when the individuals no longer act independently of but in cooperation with each other. The extreme complexity of economic life offers an abundance of examples of synergetic effects; we shall give a few typical examples, such as the behavior of businessmen.

Two Ice-Cream Vendors on the Beach Want to Do Business. But How?

An amusing example, for which I have to thank Tim Poston, one of our visiting professors, concerns two ice-cream vendors on the beach. One would have thought, naively perhaps, that it would be best for them to divide the beach in two and set up shop in the center of each half. But this state is not necessarily stable: one of the two might have the idea that he could boost his sales by moving closer to the boundary and luring some of his rival's potential customers onto his own area. The rival will react by moving closer to the boundary himself. These moves are repeated several times until both meet at the halfway line and compete strongly with each other. When Poston told me this, it seemed to us that in the final outcome both vendors would do less business than before because customers from the outlying areas of the beach would not bother to come so far. In other words, the correlated behavior of the two ice-cream vendors, in which one always reacts to the other, maneuvers both of them into a situation in which in the end they do less business than if each had kept to his original place. We find frequent examples of this situation in economics, which we shall presently discuss.

As I pondered this example a little longer, doubts did occur to me whether the ice-cream vendors would indeed always be worse off because of their actions. I remembered an observation I had made on one of my many lecture tours. When I wanted to find a particular store or restaurant in a strange city, I often searched for a long time only to find eventually that in a certain part of the city there was one restaurant next to the other and one store faced another across the street that sold the same goods. This contradicts our usual view, which leads us to expect that stores ought to keep away from competitors and set up in business spaced apart as much as possible. This led me to the assumption that the size of the catchment area, so to speak, should be an important factor in the behavior of shopkeepers; in other words, it is essential to know how mobile the customers are, how much time they have and how much inclination to cover certain distances. My assumption showed indeed that the even distribution of stores within a territory makes very good sense if individual customers want to cover only short distances. But even if they are prepared to travel further it is still better for the stores to be concentrated in a certain quarter; they will cooperatively exert a greater pull on the potential customer. The stores as a group will thus be able to offer a better overall choice and collectively to score off isolated stores. As a result the same stores that at first glance would seem to compete with each other will accumulate in a certain area. I know of businessmen who moved their relatively small stores into the neighborhood of a large shopping center because they knew that this would also increase the flow of customers to their own establishments. That this accumulation of stores of the same kind is obviously nothing new is illustrated by street names, such as Baker Street in London.

Why Do Cities Grow and Grow?

Concentration in certain places of like shops and businesses is not the only evidence of a mechanism that controls all human settlement. Certain social facilities such as schools, churches, hospitals, city halls, theaters, administrative establishments, and so on, become necessary and at the same time possible only when a settlement has reached a certain size. Here, too, the size of the settlement and the setting up of such new facilities mostly postulate each other. As communication between people increases and their expectations of let us say cultural activities grow, the wish becomes stronger to live in a settlement large enough to provide these amenities. At the same time, such settlements also hold out the promise of better economic prospects or, in poor countries, of any economic prospects at all. This may be the reason for the continued growth of large cities, emptying small towns and villages of their population or engulfing them if they are situated on the outskirts. There is an automatic increasing centralization in which a "mode," as we have seen in physics (or, in this case, a "center") becomes more and more dominant. We have a case of typical growth instability. Whether or not this growth continues depends on, among other factors, the availability of public transportation; here, too, interesting types of phase transition may occur. If the number and speed of public transportation vehicles increase no more than to enable the residents of the immediate outskirts to reach the city center within an acceptable time, cities will probably continue to spread outward. But very efficient public transportation can make the growth of satellite towns possible, some of which will have the character of so-called bedroom communities.

It can be observed, especially in the United States but also in the United Kingdom and in the Federal Republic of Germany, that the automobile plays a considerable role here. Because the prices of building land are rocketing in conurbations, people are obliged to move further out, with the advantage of living in the country. The new residential areas are often poorly developed; in particular transportation to convey the population to their jobs is infrequent if it exists at all; nor would it pay its way if it existed, because the new "settlers" are still scattered over a large territory. This leaves the automobile as the only efficient means of transportation when new areas are being developed. It is also the only means by which many people can escape from the big city. This leads at the same time to a separation of work and residential areas, with its advantages and disadvantages. The greater individual mobility also produces changes in the economic structure, as the example of the accumulation of shops and stores illustrates. Often the old-fashioned neighborhood shops no longer have a chance to establish themselves and are replaced by shopping centers with large parking lots. Once such new settlements have developed it will be important to extend traffic facilities either by the construction of roads for motor traffic or by the establishment of efficient close-haul public transportation, which may become profitable after an initial phase. As in all synergetic states of order, individual system components depend on each other for their existence: for close-haul public transportation to be economically viable it must be used by a sufficient number of commuters to make it so; but the commuters will use, say, the city railroad only if the trains run frequently enough. The initial phase will therefore always take a loss. The competition between automobile and railroad is interesting to watch. In the United States the automobile has largely replaced the train, an unpleasant surprise to many Europeans when they visit the country. Conversely, one of the first things an American will do when he arrives in Europe is to hire an automobile, to the surprise of many Europeans who would rather go by train. Obviously ideas differ widely about "how to get there."

We must be very careful before making a final judgment of all these phenomena. The automobile has provided us with mobility beyond our wildest dreams, both in our professional life and for leisure. On the other hand, the automobile creates energy problems and emits noxious exhaust fumes. In view of the most varied components of the structure of our daily lives and their multiple links, I consider it wrong either to condemn the automobile lock, stock, and barrel, or to regard it as the exclusive means of transportation. It is important to arrive at a qualified judgment and not to lose sight of the overall aspect. Often the same people who complain about a poor network of roads on a driving vacation are the ones who actively campaign against the planned building of a new road in their own community.

Unfortunately lack of space forbids dealing with these interesting problems in more detail, but the observations already made may induce the reader to think of cities and of transportation media not in terms of fixed installations but of something organically grown and still rapidly changing.

Business Management: Do What Your Rivals Do?

When we investigated the behavior of the ice-cream vendors we spotlighted a complex of problems that in economics is linked with the theory of business management. Management must conduct the affairs and evolve the best possible structure for the company and adopt the most suitable sales policy. The decisions that must be made by a company's management are obviously most varied and, at least according to current theories, left entirely to it alone to make. But the example of the ice-cream vendors shows immediately that a decision by one firm is readily influenced by that of another. One of the reasons for this is that the consequences of decisions are affected by imponderables such as the general economic situation and the attitude of the customers, for example, whether or not they accept a new product. The companies, needless to say, endeavor to minimize these imponderables through market research on the one hand and sales promotion on the other.

Here synergetic effects play an important part. When a new product is launched it will often be protected by patent. Nevertheless the establishment of the new product on the market may well be favored by the fact that it is offered by several firms simultaneously, who help each other by drawing attention to it. This synergetic effect can of course become reversed when the market moves toward saturation. We will then witness the typical behavior of systems with limited resources. We have already met many of these examples in this book, for instance in the structure of the laser modes and especially in Darwin's theory of evolution. We have seen that increased competition can be tackled in various ways: it can be countered either with progressively greater specialization, perhaps in highly prestigious products, or with a considerable spread of the product range—generalization, in brief. The automobile industry offers well-known examples. In the first case, a company would make only sports cars with a special image; in the second case, the firm would offer a varied line ranging from a subcompact car to a deluxe sedan. These indications already demonstrate that economic systems are generally not static at all. In fact, we find continuous up and down movements of the most varied events.

Because of the uncertainties of decisions, company managements are forced to watch each others' activities, which in the final resort produces a kind of collective behavior in the business world without any previous consensus having necessarily been established.

This is the right moment to point out an important aspect that is easily understood in the context of synergetics. The concept of conspiracy turns up frequently in economic theory and in sociology. It looks as if other managements or other groups (such as customers) have entered into a conspiracy against an individual. We shall see later that collective behavior produces automatic reactions that the individual cannot escape, so that it appears as if a certain group if not the whole world has conspired against him. A concrete example will demonstrate that the decision is not due to good or ill will, but to collectively evolved conditions.

Economic Boom and Slump—the Two Sides of the Coin

Here we must discuss a problem that in times of economic boom is ignored but becomes very distressing in times of economic slump—the problem of underemployment or, to call a spade a spade, unemployment. These problems obviously are of deep concern to the economists. In the past the economy was regarded as a static structure. The economists used terms such as economic viability and flexibility. How well can a firm adapt itself, for instance, to a slight change in sales opportunities? Today a dynamic evolutionary approach to economics is gaining ground. This is of course in complete agreement with the general lines of synergetics, where structures are not regarded as given entities but rather an attempt is made to understand them in the light of their origin. We shall proceed from a mathematical model consideration by Gerhard Mensch, which is easy to incorporate into the general processes of synergetics.

We are all familiar with the fact, corroborated by economists such as Haberler and many others, that industrial development passes through phases of boom and slump, with sometimes quite pronounced transitions. The numerous examples given in preceding chapters have illustrated that in many systems even minute changes of the environmental conditions, which are called "controls," may produce drastic changes in the general structure. Let us investigate the question of full employment in the light of these experiences. Before we begin to trace the underlying causes of these phase transitions we must mention some relevant observations arising from empirical economics research.

Are Technical Innovations Always the Driving Force of the Economy?

We have seen time and time again that two widely different areas exist in the behavior of the most varied systems. In one area, for example, a lamp or a layer of liquid behaves normally—i.e., to all practical purposes it continues its behavior—if the disturbance is minor. But we also have the highly interesting areas in which a system becomes unstable and wants to assume a new state; conditions have come as it were to favor a transition to a new state. The time of the new state's appearance and its mode are often triggered by random fluctuations. We find this very behavior within the economic models we are discussing. But what assumes the role of the fluctuations in the life of the economy, the role, so to speak, of the triggering moment? A group of events in economics consists of innovations, especially of those based on inventions, such as the internal combustion engine, the airplane, the telephone, even a new vacuum cleaner. A large group of inventions that although less noticeable is nevertheless also very important concerns the simplification of production. The economist calls them innovations.

According to the empirical observations of innovation research, a first phase begins with basic innovations, which establish new branches of industry. A dramatic example is the invention of the automobile. Such basic innovations usually appear in large numbers, that is, they accumulate; they are followed by those innovations that have the purpose of improving production in the newly established branches of the economy. The expansion of one such branch favorably affects the other branches so that a general economic boom is the result. This happens in many different ways, for instance through high employment, which generates high purchasing power, which involves the inclusion of subcontractors, and so on. As economic investigations have further shown, the innovations that permitted the manufacture of *new products* by far exceeded the introduction of *new manufacturing processes* in the industrial countries of Europe during the late forties and throughout the fifties. In the sixties, the innovations were displaced: the production methods changed, a change that can largely be defined by the word "rationalization."

The overriding incentive for economic activities is without doubt the profit motive. A discussion of this is often quite emotive, as when a motorist thinks of the rising prices of gasoline and the profits someone reaps from it. But let us ignore emotions here and bear in mind that reduced profits eventually turn into losses, and such questions as job security often become acute. Let us confine ourselves to the economic aspects only: on the one hand, profits involve the sale of a sufficient volume of products, but on the other hand, a company's profits are reduced for instance by higher wages, which affect prices and may make competition difficult. At the same time, expansion of production is often linked to the introduction of new products, which initially is expensive. Both—higher wages and the avoidance of high initial outlay—lead to investment not in expansion, which is to say increased sales, but in rationalization. Consequently companies prefer innovations that result in an improvement of the production processes rather than in new products. An automobile manufacturer will prefer introducing a new automatic welding machine to a completely new model of vehicle.

As previously mentioned, G. Mensch developed a mathematical model derived from the so-called catastrophe theory with the aid of empirical data. It describes the transition to be observed from full employment to underemployment. I have translated this model into the language of synergetics and expanded it. Examples such as the laser and liquid rolls show that we can read almost directly on a graph what new positions of equilibrium will be established when we change external conditions. Figure 12.1 shows the progress of synergetic input when we change the production called x. We assume an economy that is in equilibrium, to start with. We must now investigate how the synergetic

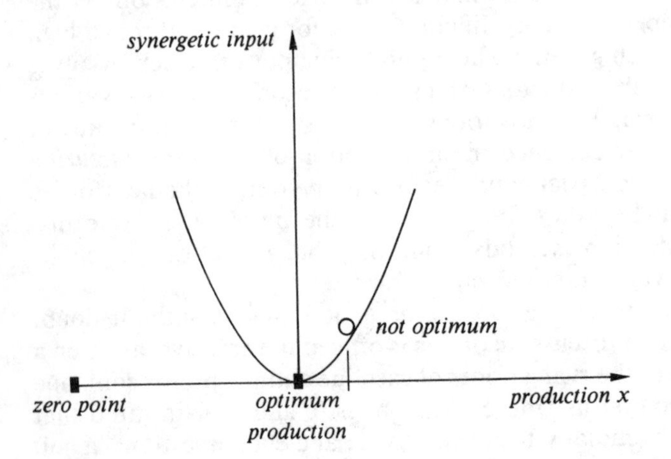

FIGURE 12.1 Graph of the synergetic input when production of magnitude x is changed.

input changes when we introduce changes in production. Because the equilibrium is stable initially, we find the curve shown in figure 12.1

We now have to examine the change in the course of the input curve resulting from investment in an increase of production. Because this is our aim the curve must, needless to say, be shifted to higher production levels; i.e., a curve as shown in figure 12.2 must be achieved. Conversely, measures that reduce production shift the curve to the left. Now let us look at what happens when companies rationalize—i.e., streamline their production methods. This may have two entirely different effects, as some examples show. New machinery replaces manpower; this lowers labor costs, which enables the company to make a profit even when its production is reduced. Here rationalization is designed for lower production. Or it may be designed to make the products cheaper, and the increased production is absorbed by the market. Both the possibilities of reduced and of increased production can be represented by the synergetic curve of figure 12.3 and are familiar to the reader. Our illustration shows without a doubt that we must abandon the theory of *only one* possible equilibrium of the economy. Here two stable states that are fully equivalent from a purely economic aspect are possible. Stability means, after all, that the states do not undergo major changes when we disturb them extraneously. In this course of events the attempt could obviously be made to break the symmetry from the outset by external interference, through the introduction of artificial differences into the synergetic input as shown in figure 12.4. We shall explain below how this can be achieved.

But a factor peculiar to economics complicates matters: only a limited amount of investment capital is available for the introduction of innovations. Reduction of investment shifts the curve in the direction of reduced production. A combination of rationalization methods and reduced investment results in the picture shown in figure 12.5, which indicates that the economic situation is clearly determined by the minimum on the left; that is, lower production and indirectly also lower employment.

This representation illustrates a conclusion reached by G. Mensch. To take advantage of rationalization for increased production and thus full employment, investments would have to be made that lead to increased production in order to realize the curve of figure 12.4 (fig. 12.6). But the higher production will be absorbed by the market only when it is combined with innovations that result in novel products.

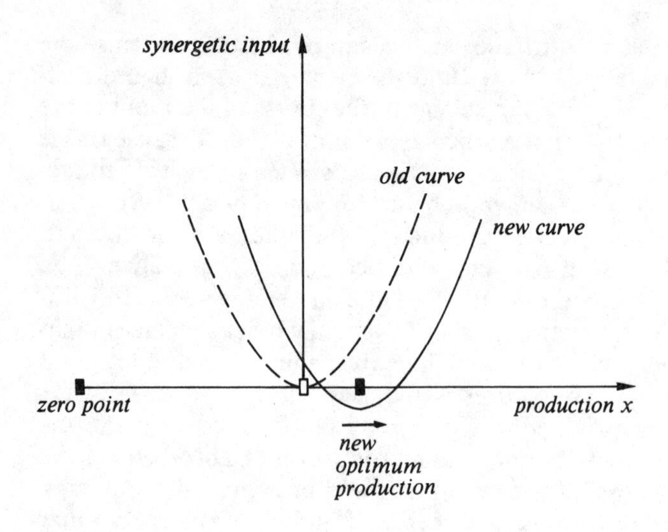

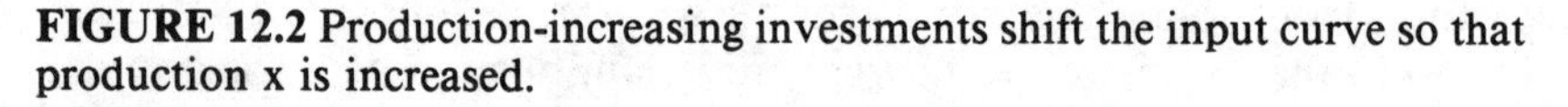

FIGURE 12.2 Production-increasing investments shift the input curve so that production x is increased.

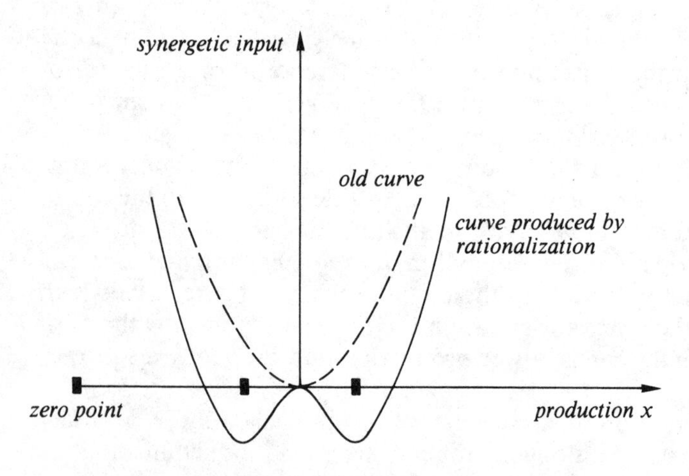

FIGURE 12.3 Rationalization measures produce two optimum positions of production: either increased or reduced production.

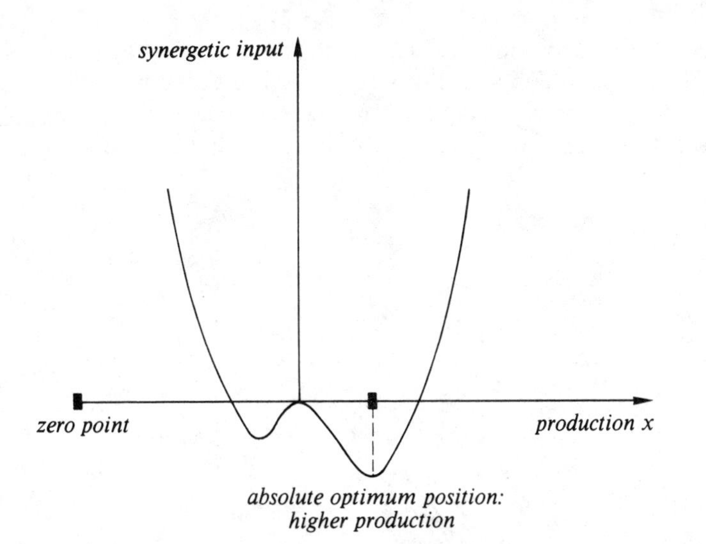

FIGURE 12.4 The joint effect of rationalization measures and production-increasing investment achieves an absolute optimum position, linked with higher production.

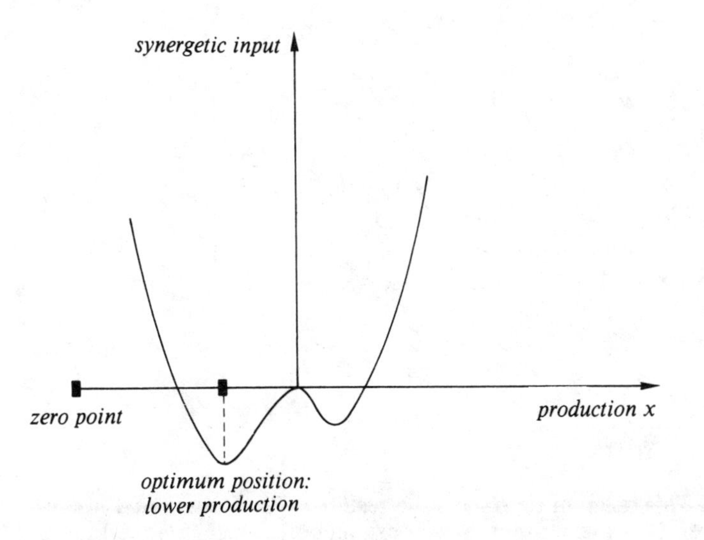

FIGURE 12.5 With reduced production-increasing investment but with rationalization measures, a position will become more favorable to companies in which production x is cut back.

I considered these fairly technical explanations necessary to show the reader how very complex situations can be represented relatively simply by the ideas of synergetics and how their effects become very markedly evident. On the other hand, as in all the other chapters we

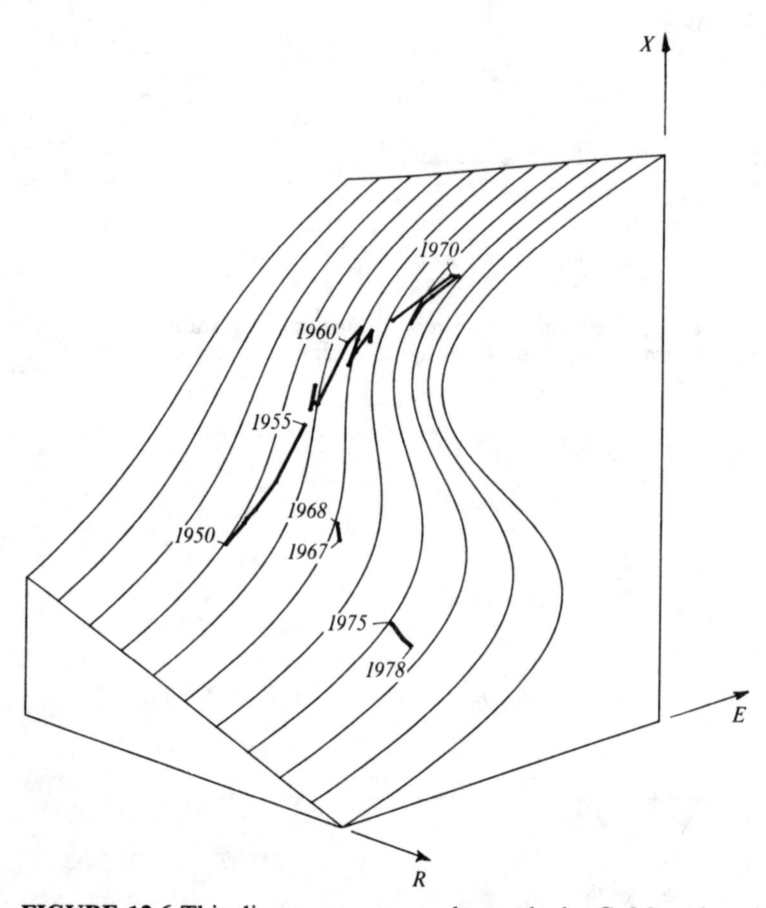

FIGURE 12.6 This diagram represents the results by G. Mensch and his co-workers. It shows how the optimum size of production changes as investments for rationalization (R) and expansion (E) are made. The years refer to production in the Federal Republic of Germany. Note the marked drops to underproduction.

must be fully aware that a complete theory of such events would obviously take up a whole volume, which is quite beyond the scope of our discussion; all I want to do is stimulate further thought and point out how models can be set up of even complex reactions.

We must be aware that, in addition to the mathematical treatment of the problems, the interpretation of the conditions and the results also plays its part. We thus mentioned in the context of reduced profits that higher wages may also be responsible for this reduction. Conversely, at least in part higher wages are based on price increases, which in turn are the result of increased production costs and therefore of higher wages. This is the wage/price spiral familiar to all of us. But as one phenomenon is the cause of the other, from the point of view of synergetics it is rather futile to look for the culprit. What we must appreciate is that the wage/price spiral, for example, causes a change in the parameters, which themselves may result in the abrupt turnaround of economic trends we have just discussed.

Abrupt, Collective Changes in Economic Life

A comparison of the synergetic input curve with empirical data shows that the economy is obviously able to sense the development of a lower minimum of this input curve and to react to it by "jumping" into this new minimum (fig. 12.7). It is interesting to note, though, that this action is often delayed. Economists are frequently at a loss about the causes of this "jump." Usually they look for external causes such as an increase of the price of oil, but internal causes are more likely to be responsible for the previously discussed attitudes of companies toward investment. The economic situation had changed to the extent that this jump was already long overdue, but nobody dared make it. The effects are the same as those that occur in the physical arena when water is undercooled. The water temperature has dropped below the freezing point, and the water should long have turned into ice. It is in what is known as a metastable state and does not freeze. A spontaneous fluctuation or a very slight external shock makes the water freeze abruptly.

The situation is similar in economics. The processes may very well have an internal cause—a *single* company may decide to take measures with a view to greater rationalization, for example. But the time

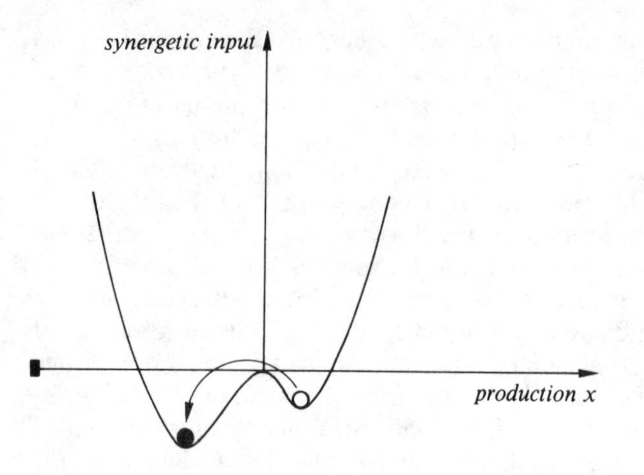

FIGURE 12.7 Jump of the economy from a relatively favorable position to an absolutely favorable position for profits or synergetic input of the companies, but not for the general employment level.

for this is already overripe, and all the other companies follow suit. The action of the first company resembled a fluctuation and had the effect of a signal. It seems that such lines of action are often disguised with the excuse of "external" causes. But these are much too unimportant to justify such far-reaching new decisions by themselves. The root cause was the *general* economic situation, which now takes on the part of "control parameter." These "external" causes merely provide the impact that makes the ball of figure 4.16 roll down the synergetic input curve, but the "internal" causes—i.e., the general economic situation—had long ensured a distortion of this curve. The ball is no longer situated in the "correct" stable spot. A strict division into external causes (or impacts) and internal ones is not always possible. A continued rise in the price of oil may well lead to a systematic "distortion" of the synergetic input curve and thus bring about a totally changed economic situation; in such a case the price of oil itself has become a control parameter.

These observations also explain why many companies in the same business often act very similarly within a very short time. If one of them diverges from this mode of action, it would act against the general economic trend and thereby place itself on a most unfavorable point on the synergetic input curve.

It is clear that seemingly collective decisions by companies are not necessarily based on mutual agreement, which is of course quickly suspected to contravene antitrust laws. On the other hand we do not claim that this never happens. But rationalization (and other measures) need not *only* be a consequence of the economic situation. Firms may also rationalize their methods *in expectation* of a shrinking market, of a toughening competition. Rationalization can itself lead to a shrinking market, because employees are let go and general purchasing power is reduced. Cause and effect can no longer be distinguished.

This also demonstrates that the question of how to lead the economy out of an underemployment equilibrium does not necessarily have a clearcut answer, but also depends on complex processes. G. Mensch's conclusion—increased investment that has the effect of expanding production will result in the other stable state of full employment—is not inevitable, as for example when the market cannot absorb the increased production. The economy can therefore also be stimulated via an increase in the purchasing power, which can be brought about for instance by a reduction in taxation.

You will have noticed that we have become deeply involved in economic theory; but you are also familiar with the basic principle that even minor changes in circumstances can produce radically different states of equilibrium.

Economics Is Not as Simple as Adam Smith Thought

Even at this stage our reflections and observations contradict dogmas of the traditional free market economic theory traced back to Adam Smith, who proceeded from the assumption that with untrammeled competition an equilibrium, generally thought to be the only one, will always be established. But we have already discovered the counterexample of a case of two possible states of equilibrium. Here the general economy finds it extremely difficult to manage the jump from one equilibrium to the other, because this is primarily possible only by common action. In reality, economic behavior patterns are far more complicated. It is, for instance, possible for an economy to fluctuate continually between the states of equilibrium. Full employment periodically alternates with underemployment.

Certain automatic reactions, then, take place that may equally well result in undesirable phenomena such as unemployment; they immediately raise the question of whether the introduction of external—i.e., government—controls will be able to prevent such effects. But we must be aware that controls cover a very broad spectrum of the most varied measures. An example of a concrete physical system is the laser. We have learned that by changing a single environmental parameter—i.e., the intensity of the current supplied—we can induce the atoms to organize themselves for the emission of coherent light. We imposed a highly nonspecific control here, which affected all atoms equally but nevertheless produced a very detailed orderly behavior of the laser. Another possibility exists: with the aid of special fields of light, for instance, we can so control each of the atoms externally that they all emit at the same rhythm. This obviously requires an enormous amount of force, because we would have to target and control *every single* atom directly. In the field of economics the possibility of control on the one hand and how this affects the economy on the other produces a quite similar situation. As models immediately show, it requires an enormous effort to check and control individual events, so that the controls would cost more than would be saved by, for instance, the balancing of events. This is an experience to which many government authorities and especially the bureaucrats refuse to open their eyes.

The ingenious feature of the laser is the very possibility that we can with very little effort, without any detailed information about the state of the laser, use a very simple measure to induce the laser atoms to organize themselves. Nor does there exist any doubt at all that particularly the perceptive economic experts support the point of view that the economy requires the *least complicated* controls. It is unfortunately only too obvious that the government inundates us with a flood of the most varied controls in the form of a wide range of different taxes and detailed laws on the one hand and selective subsidies and indiscriminately distributed benefits on the other. Let us mention two examples, of which the second is political dynamite.

Because of the devastation of the Second World War, the German government was faced with the task of supporting housing construction. It is interesting to note that this is one of the fields in which we find the two just-discussed possibilities of controls put into practice. In one possibility, the government supplies all the funds to finance housing construction. In the other possibility, there is a control parameter that in itself does not carry much financial weight but guides the flow of capital in the right direction. This control parameter consists of tax concessions for private individuals who want to build houses. Their capital investment is thus channeled in the desired direction without the need for the government, which is to say the public, having to find all the capital for this purpose.

The second and politically highly inflammatory control consists of the rent restriction acts, which above all meet the social need to "protect the tenant against eviction." The legislation is designed to freeze rents—which can be easily seen, but also mathematically proved—at a certain level. But at the same time that this protects the tenant, it also robs the landlord of the incentive to build new houses, because these will very quickly become unremunerative. The end result is a considerable housing shortage because private investors have diverted their money into other apparently more profitable branches of the economy.

This is a clearcut example of a conflict situation in the sense of an alternative: the legislator must decide in favor of one or the other case; priorities must be established. We recognize clearly how legislation has a direct bearing on economic processes even when this had not been its original intention at all.

Expectations are now probably high among many readers that synergetics offers a cure-all for such difficulties. This is by no means so, but not because synergetics is not yet sufficiently developed; quite the contrary. We have seen in innumerable synergetic examples basic conflict situations in which one solution precludes another. The only possible remedy is to soften conflicts by means of greater differentiation. But this too may involve such great external effort as to be no longer worthwhile. Here is one more example of government intervention that may have far-reaching effects. The fact that even small changes of environmental conditions may cause drastic changes in the system as a whole must by now be familiar to the reader. Some such environmental condition, which here appears as one of the facts of life, is the already mentioned taxation. Situations may very easily arise in which even a very slight increase in taxation can drastically change the spending habits of the population, with the possible result of a very rapid development of entirely new macroeconomic conditions such as increasing unemployment. It is my personal view that quite a large number of politicians are still blind to the possibility that small changes in the environment (equal to changes in the very facts of life) may bring about drastic changes in the state of a whole system.

Economic Chaos through Ill-Considered Controls

We must mention one more point, one which is absolute anathema to many economic experts; but it is mathematically based and will without doubt become established in economic theory within the foreseeable future. Examples from physics and chemistry have taught us that even controlled reactions may proceed chaotically. Let us envision a chemical reaction that progresses periodically, complete with a color change of the substance from red to blue to red, and so on. One can of course say that the color change is too slow, and to control the process a substance is added periodically to speed it up. It has been shown both experimentally and theoretically that the behavior of the system may change completely. A regular, periodic color change is replaced by an entirely irregular, chaotic one.

The situation is exactly the same in the highly complex systems of the economy. We must in fact expect that control measures that ignore the peculiarities of a system can lead to typically chaotic processes. A whole literature on chaotic behavior exists in the natural sciences as well as in biology, and economists would be well advised to familiarize themselves with this complex of problems.

Does Peace Become More Secure through Closer Economic Ties?

We have seen that a large number of phenomena from economics can be considered analogous to those of physical systems. This is due to the fact that at least to a certain extent the economy can be described with the aid of mathematical laws, and that the analogies in the results follow because of the similarities of the mathematical relations. From this aspect we have looked at models that deal with the following question of considerable immediate interest: is global peace made safer by closer economic ties?

Powerful tendencies exist in many countries that aim at forging close economic links, especially between opposing political systems, with the object of thereby making global peace more secure. The mathematical formulation of such events produced a result that at first surprised me greatly: it was shown that increased stability can indeed be created through closer ties; but it was also shown that a state that had been stable before the establishment of such ties may in fact suddenly become unstable and a catastrophe can ensue.

The interpretation of increased stability has now almost become common political knowledge. Each of the partners sees that he will profit by close economic ties, and therefore far from wanting to endanger the relationship wants to strengthen it further if possible.

Does the mathematical model fail in the second case, that of instability, or are there profound reasons for it to develop?

Instabilities become effective only as a consequence of fluctuations, as we have found again and again. Such fluctuations in the life of nations, for instance, may be due to various crises, economic, political, or military; some of them may be quite local. But such fluctuations may lead to measures by the other side which may take the form of economic reprisals; these are then answered in kind, resulting in an explosive development of the conflict.

This model example shows at least that closer economic links do not automatically increase political stability.

It appears in fact necessary to build this stability on a more solid basis than economic links; this goes beyond the purely mathematical aspects, and only increased mutual trust will bring it about.

Synergetic Laws Benefit Mankind

We have given a number of concrete examples of often striking similarities between economic processes and those in entirely different areas such as physics and chemistry. Here collective behavior plays a decisive part. But phenomena may occur on this basis that diverge radically from Adam Smith's classical economic theory postulating an equilibrium. Without any doubt economic theories of the future must concern themselves in depth with these novel phenomena and the methods of synergetics, to be able thereby to create a better understanding and even to improve the functioning of the economy. But in this context we must not overlook the following circumstance. Like any other theory of economics and especially of the social sciences, synergetics is confronted with the problem of the interpretation of its mathematical results. The particular reason for this is that all economic processes have far-reaching social implications; they deeply affect the life of every individual in the professional as well as in the private sphere. This sometimes results in the a priori rejection of the mathematical approach. We encounter the usually derogatory expression of "technocrats," a species perhaps regarded with disfavor because the conclusions they reach sometimes contradict ideological wishful thinking. We must however, be aware that there are inevitable developments in a large number of processes in complex systems, economics among them, which we cannot escape through ideological wishful thinking. Rather we must find out what course these automatic events take so that we can make use of them on a higher plane for the benefit of every single individual.

Chapter 13

Are Revolutions Predictable?

In his science fiction series, *Foundation*, Isaac Asimov describes a scientist, Dr. Seldon, who was able to calculate for many centuries in advance the behavior of the masses, which put him in a position particularly to predict revolutions. This is a question of obvious interest not only to futurologists. It would be very useful to the ordinary citizen, let alone to the politicians, if we could make such a prediction even by only a few years.

In a revolution one political order is replaced by another through force. The word "order" points directly at the central question of synergetics. How is order established through the cooperation of the constituents of a system? Applied to this chapter, how is a political order established through the cooperation between and among the individual citizens?

Public Opinion as Order Parameter

Once again we are faced with the peculiar interrelation between individuals and the state of order; this enslaves the individuals, who in turn uphold it. Let us look at this interrelation in greater detail in the example of sociology's burning subject, the formation of "public opinion."

This is our theory: the prevailing public opinion plays the role of the order parameter which enslaves the personal opinions of the individual, that is, it enforces a largely conformist public opinion, thereby keeping itself in existence. The thesis that the opinion of the individual is enslaved must of course be reasoned in detail, and I shall show that an abundance of positive evidence is contained in the sociological literature. Conditions in the area of the sociological, however, are more complex than those in the laser or in the liquid, because here additional forces are at work, as it were other part-systems—the mass media on the one hand and the government on the other. It will nevertheless quickly become apparent that the concepts of synergetics enable us to clear a path through the jungle of the various nexuses, offering us a fairly lucid picture of the interrelations between the separate constituents of a society.

The theses we present are as follows:

1. Individuals can be influenced by a prevailing opinion and tend to follow it.

2. There are basically two ways open to individuals to learn about the opinion of others: by direct contact with each other or through the mass media.

3. The mass media have a dynamism of their own.

4. Among the mass media, the press is open to the collective influence of the citizens through the degree of its popularity among the public.

5. In a democracy the government is decisively shaped by public opinion.

It has become possible within the framework of synergetics to treat a number of these interrelations between order parameter and enslaved system by means of mathematical models and thus to simulate the dynamics of opinion formation. Let us examine how the concept of the order parameter on the one hand and enslavement on the other enables us to illuminate the various interrelations among the different forces within society; figure 13.1 is a diagrammatic representation of the situation. The arrows denote influences that we must investigate in some detail.

The claim that the order parameter enslaves the subsystems acquires a shocking dimension in sociology, because it indicates that the formation of the individual's opinion is enslaved by the prevailing opinion (fig. 13.2). This claim is so provocative that one instantly tries to reject

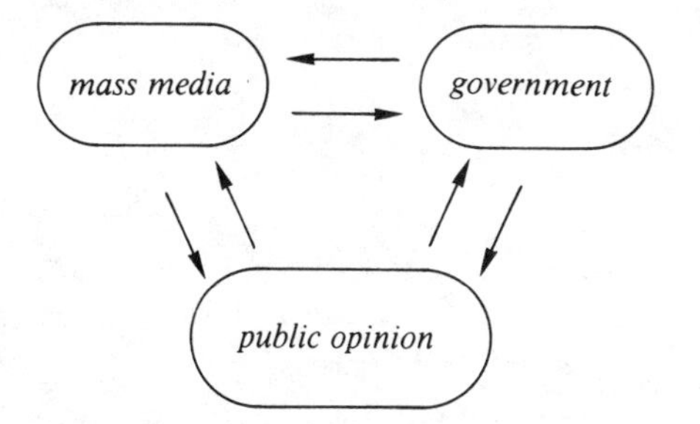

FIGURE 13.1 The interrelation between government, the mass media, and public opinion.

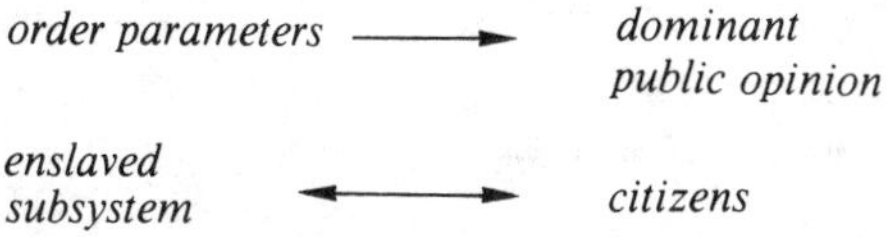

order parameters ⟶ *dominant public opinion*

enslaved subsystem ⟷ *citizens*

FIGURE 13.2 Opinion formation by the individual is enslaved by the dominant public opinion.

it as an inadmissible extrapolation from the natural sciences to sociology. But we must not be too rash with our judgment; let us ask the sociologists themselves. In her book, *Die Schweigespirale* (The Spiral of Silence), the well-known opinion researcher Elisabeth Noelle-Neumann has collated the observations of leading sociologists, and they confirm this thesis. The numerous examples in her book of ever new kinds of order parameter and enslaved subsystem prompt the question of how we can precisely formulate the concept of the order parameter. Because we want to identify this with the dominant political opinion, we must therefore ask ourselves what is "public opinion." Dozens of definitions of this term are to be found in the sociological literature. Since the basic approach of this book is the natural-scientific one, we must confine ourselves as much as we can to dealing with measurable quantities and exclude ill-defined ideas. We can directly adopt the procedures of the opinion research institutes.

These institutes compile a list of questions on topics of current interest. Here are typical examples: "Are you in favor of the death penalty?"

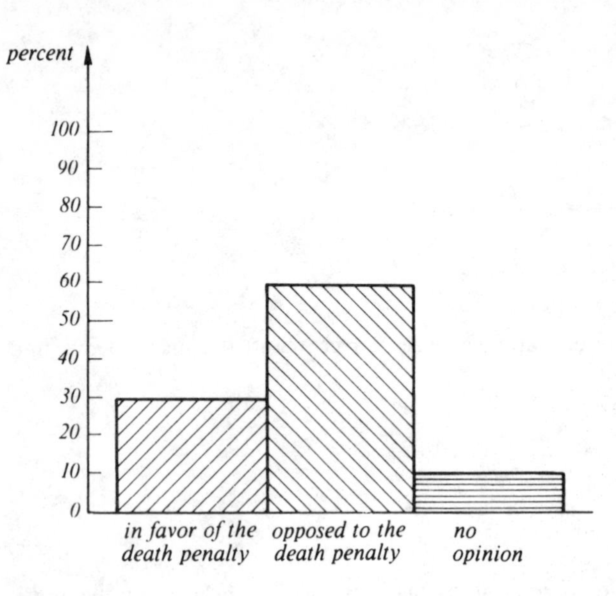

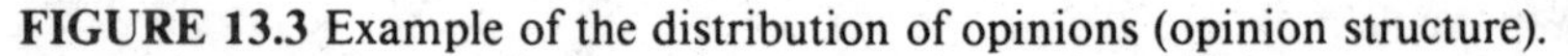

FIGURE 13.3 Example of the distribution of opinions (opinion structure).

"Which party would you vote for if there were a general election today?" And so on. The pollsters count the votes for both parties, which gives them a picture of the distribution of opinion. Such a diagram shows immediately which political opinion is dominant. This is the equivalent of the investigation of a structure in the scientific tradition (fig. 13.3).

But synergetics is about the understanding of how structures *come about*—i.e., about dynamics. We must therefore ask ourselves how a particular distribution of political opinion originates. We can think of two extreme cases. That of the completely mature citizen who independently forms his or her own political judgment and fully stands by it, and that of the man or woman who when forming an opinion allows him- or herself to be influenced by the opinion of others. In the second case it is the interrelations or, in other words, the synergetic effects that play a decisive role.

Can People Be Influenced?

Sociological material reveals that we must always expect people to be influenced when forming their opinions. This is caused by the psychological makeup of human nature, which we are about to investigate; it is also a natural reaction to the environment. Modern civilization has created an extremely complex human environment in which people do not always find it easy to orient themselves. Each individual is pushed from one conflict situation to the next, where it is difficult to produce a clearcut answer by his or her own efforts. This generates the tendency to look at the actions and opinions of fellow citizens. Furthermore, some experiments by social psychologists reveal that quite an appreciable percentage of persons will follow a trend of opinion that they ought to, and perhaps even will, recognize as objectively wrong. Probably the most striking example is based on experiments by the American social psychologist Solomon E. Asch, lucidly described by Elisabeth Noelle-Neumann in her previously mentioned book:

> At the beginning of the fifties a report was published in the United States about an experiment performed more than fifty times by the social psychologist Solomon Asch. Volunteers had to estimate the length of various lines as a ratio to a reference line. Of three given lines one always had the same length as the reference line [fig. 13.4]. It was an easy task—or so it appeared at first glance—because the line in agreement could be readily identified. Eight to nine persons participated in the experiment, which proceeded as follows: as soon as the three comparison lines were suspended next to the reference line, each person, in the sequence from left to right, indicated which of the three lines corresponded in his or her opin-

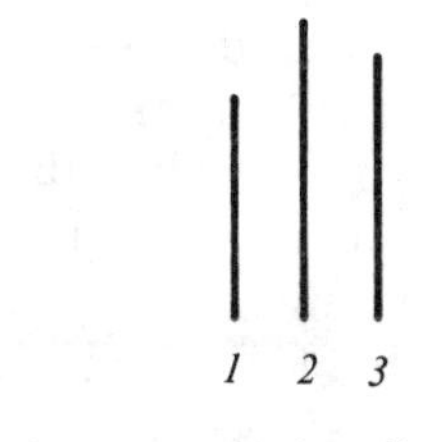

FIGURE 13.4 The Asch experiment.

ion to the reference line. Each experiment had to be performed twelve times, with twelve repetitions.

The following test was now carried out: after all the volunteers had agreed on the correct line during the first two runs, the experimenter changed the situation. His assistants, who knew the purpose of the experiment, all indicated too short a line as being in agreement with the reference line. The behavior of a naive volunteer, the only person unaware, who sat at the end of the row, was now examined under the pressure of a dominant different opinion. Would he begin to waver? Would he agree with the majority verdict, no matter how much it contradicted his own? Or would he uphold his own judgment?

Result: of ten volunteers, two could not be persuaded to change their minds; two agreed only once or twice during ten test runs, but six out of ten indicated several times the obviously wrong minority view as their own. This means that even in a harmless question and in a rather indifferent situation that does not affect their real interests most people follow the opinion of the majority even when they can have no doubt that it is incorrect.

In synergetic terms, this yielding to influences is the root of all collective effects in the formation of public opinion. It is quite immaterial how this influence is exerted and even how we formulate it mathematically. According to the general laws that are the basis of synergetics a competition automatically arises between the various opinions, with one coming to dominate and eventually winning the contest. A particularly striking example of this phenomenon is fashion, which is no more than another manifestation of public opinion. Here indeed we have a very spectacular demonstration that in such collective effects objective yardsticks are often unimportant, and a subjective tendency is ultimately preferred in the collective. Whether women wear long or short skirts and men tight or wide trousers is exclusively a question of taste, not a result of profound deliberations (unless the fashion designers want to stimulate business and cleverly succeed in exploiting the collective behavior of people for their own ends). This applies not only to the fashion designers but it is also probably the secret of the successful politician who knows how to take advantage of such collective trends. We shall presently return to this phenomenon.

If the synergetic effects are really active in the formation of opinion, it would obviously be most surprising if they had escaped the notice of the sociologists. Far from it; there is abundant evidence of their awareness of them in the sociological literature and historiography. The concept of public opinion, *opinion publique*, may first have been used by Jean-Jacques Rousseau (1712–78) in the sense that it is a pronouncement of judgment, whose disapproval should be avoided. In the terms of our current definition it would be preferable to speak of a "dominant public opinion." That man is swayed by influences was clearly expressed by James Madison (1751–1836), one of the fathers of the American Constitution, when he observed:

> If it be true that all governments rest on opinion, it is no less true that the strength of opinion in each individual, and its practical influence on conduct, depend much on the number which he supposes to have entertained the same opinion. The reason of man, like man himself, is timid and cautious when left alone, and acquires firmness and confidence in proportion to the number with which it is associated.

Elisabeth Noelle-Neumann defines this relation as follows: "His social nature causes man to fear isolation, to want to be highly regarded and liked by his fellow men."

Elsewhere she says:

> Only if we assume a very deep fear of isolation can we explain the enormous human achievement at least in the collective when people can say with great accuracy and without any demoscopic aids what opinons are on the increase and what on the decrease.
>
> The effort of watching the human environment is seemingly the lesser evil compared with the risk of losing the goodwill of one's fellow men, of suddenly being isolated.

Alexis de Tocqueville (1805–1859) recognized this human openness to influence much earlier: he formulated it as follows:

> In the democratic nations public favor appears to be as necessary as the air we breathe, and to be at odds with the masses is tantamount to not living. The masses do not have to resort to laws to subjugate the dissidents. Disapproval is enough. The feeling of their isolation and impotence soon overwhelms them and deprives them of all hope.

The possibility of being influenced obviously does not yet say anything about the establishment of a state of macro-order, in our context that an opinion ultimately becomes dominant. This can be proved only with the mathematical methods of synergetics. The possibility of mutual influence results in the same amplification effect as we have found in the laser light. If a certain wave is dominant, it will ultimately win the competition with all the others. More and more atoms will be enslaved by it. This can be compared with the formation of a dominant public opinon. More and more people are engulfed by it and in the end support it.

De Tocqueville also clearly recognized the ruling principle of synergetics, enslavement. He stressed how the democratic nations first overcame those forces that "totally obstructed or delayed the rise of individual reason," how they blazed a trial for intellectual freedom. But if "on the strength of certain laws"—de Tocqueville means the authority of the numerical majority—"intellectual freedom were to be strangled, the evil would merely have changed its face; man would not have found the means to an independent life; he would have discovered merely a new form of servitude."

James Bryce (1838–1922) described this situation even more succinctly in an article which speaks of the tyranny of the majority.

It is therefore not surprising that what we have called the *dominant* public opinion as the final result of these processes is identified simply as *public opinion* pure and simple by researchers, especially by Elisabeth Noelle-Neumann. She accordingly defines as "public opinion those views in the controversial range that may be publicly held without the risk of isolation." Dominant public opinion and the opinion of the individual cause and stabilize each other in the sense of synergetics.

Change of Opinion: How and by What Means?

How can a change of opinion take place at all? Here, too, the analogy with the phenomena of the natural sciences is useful. We saw that rolls form in a liquid when we increase the difference in temperature. Applied in the field of sociology this means that changes in the environment—for instance in the economic situation, or the appearance of overwhelming pressures in internal politics, and so on—may undermine reliance on an existing opinion; in other words the system will become destabilized. This also includes acts of terrorism, which are attempts to weaken and to shatter confidence in the order of society, in justice, and so forth, in order to prepare for a change of opinion. Especially in times of radical upheaval the individual finds it highly important to watch the behavior of his fellow men to avoid isolation in changing circumstances. According to synergetics, changes in opinion are prepared by external conditions. But once destabilization has set in the view spreads among the public that something new must be done. The question of the direction in which we should proceed is, however, often in the balance. The new direction is always decided by only a few individuals. A single group of people, of avantgardists or of active revolutionaries, indeed often a single personality can become a starting point for a new direction. It is the fluctuations that are decisive, and that we have come across time and time again in this book. During a state of instability, unpredictable, seemingly local events acquire an enormous power of attraction of which they would be utterly devoid in normal times, when the actions of such isolated groups would have been a soon-forgotten episode, a minor fluctuation quickly subsided.

The Mass Media: Order Parameter under Selective Pressure

We have up to now pretended that the occurrence of an order parameter—i.e., "prevailing public opinion"—and the opinion of the individual citizen are a closed cycle, such as the laser atoms generating the laser light, which in turn enslaves them. Although this basic idea is entirely correct, it calls for an important addition: public opinion is created not only by direct contact between people, but also through the mass media. It would be naive to claim that the mass medua are but a mere replica of public opinion. That is a widely held but mistaken assumption. In fact, the media have dynamics of their own, and that is closely linked to the fundamental questions of synergetics.

The French writer Guy de Maupassant (1850–1893) not only wrote novels of piquant amorous adventures, he was also an astute and critical observer of his age. Having himself been a journalist at one time, he described in his novel *Bel Ami* a publisher who at supper assesses the incoming news for its value, like merchandise. This presents us with a number of principles the readers of this book have long been familiar with. One of them is the problem of limited facilities. A newspaper cannot publish everything; it must not exceed a certain volume if it is not to become too expensive. In addition, readers only have a limited amount of time available for reading, as the sociologists have found out—about fifteen minutes per day. The journalists are therefore obliged to treat their material selectively. But what are the criteria of selection? Obviously there are quite a number. Let us take those that are most convincing from the aspect of synergetics.

Newspapers, magazines, journals can exist only if they are bought by the public, in other words if they are sustained by their readers. But because this sustenance is limited, competition must arise, which results in a process of selection. A newspaper or periodical is therefore compelled, for reasons both of competition and lack of space, to select its material in a way that is the best guarantee of their continued existence. This creates a dual role for these types of publication. One aspect is this: they function as order parameters in that they are able to influence the opinions of their readers through the opinions they propound. According to Elisabeth Noelle-Neumann, this influence is sometimes felt to be bullying and oppressive: "The mass media are the embodiment of the general public, wide-ranging, anonymous, unassailable, and impossible to influence."

Elsewhere she has this to say:

> The mass media are one-way, indirect, public communication, they are the opposite of the most natural human communication, the personal conversation. It is this that induces in the individual a feeling of impotence in the face of the mass media; every public opinion survey that tries to establish who has too much power in our society today rates the mass media among the top.

The impotence of the individual is revealed in two ways. He or she is deprived of publicity, in other words the individual is unable to communicate his or her opinion to others through the media; the individual can be pilloried without an adequate means of defense. Even court action against a newspaper or periodical can add to its publicity, promote its sales, and thus benefit it even if it eventually loses the action. The other aspect of this dual role is this:

As much as it is correct that the individual is helpless against the mass media, these too are vulnerable and can even be killed outright by the collective behavior of their readership. I cannot imagine any newspaper successfully surviving for any length of time without external subsidy if it consistently represents an opinion that flies in the face of that of its readers. There is, however, one proviso: these publications are not confined to purely political features (nor is the political situation necessarily the main concern of a newspaper or a periodical). Simply to protect themselves against the whims and reactions of the public, many of them have taken the road to generalization. They offer a relatively broad spectrum of the most varied aspects of politics, the economy, the arts, and so on. In addition local news, which may be quite trivial, such as the time of the next garbage collection or a list of events, contributes to creating a faithful readership. On the other hand these features are not exclusive to any one newspaper and therefore do not prevent local competition. This is borne out by the observation that in small localities often only a single newspaper survives, or that in the catchment area of large cities the national news is provided by a central editorial office of one newspaper, and the individual local papers merely add their local section.

It must be accepted that an ideological competition has ceased to exist among newspapers, and that the main newspaper has acquired a monopoly of opinion; this can hardly be broken even economically by the collective behavior of the readership, which cannot or does not want to miss the local news. Here it is quite possible that in the course of the affected newspapers' development, initial preference of minor trends will have been increasingly amplified through the frequently mentioned feedback mechanism. It is an interesting task for sociologists to find out whether the election results in the Federal Republic of Germany, which have an evident north-south differentiation, are not linked to such mechanisms.

How can we extricate ourselves as individual citizens from this seemingly almost inescapable enslavement? We can overcome our own susceptibility only when we permit the external influences to cancel each other out, like being caught in a shoving and pushing crowd and managing to remain upright because we are equally pushed and pulled from all sides. We can bring this balance about only by also reading national and, if possible, foreign papers of various persuasions. This obviously does not mean that we should subscribe to twenty different papers. It is quite enough to browse occasionally in one or another. Many readers will have had an experience similar to mine when traveling abroad: problems in the United States or in the United Kingdom will appear in a completely different light and context.

Newspapers, and to an even greater extent periodicals, not only generalize but also specialize, for instance by addressing themselves to special groups of readers. There are newspapers with quality contents that are an intellectual pleasure to read, and others that seek to curry favor with their readers by descending to the lowest common denominator.

Whether or not a newspaper is read is decided not only by its contents, but at least to a certain extent also by its price. Here, too, effects are at work that lead to the preference of a single newspaper. If its circulation is higher, it can become cheaper for obvious economic reasons. This will further increase its circulation, and the process can go so far that only this one newspaper will survive the competition. Although you may well agree with the political opinion of the newspaper that has won the competition you will automatically have exposed yourself to a monopoly of opinions. Even if you wanted to stick to your own opinion (which is by no means always a sign of intelligence any more than a change of opinion necessarily indicates a lack of character), the newspaper too may in the course of time undergo a change of opinion, and you will be unaware that you are being enslaved by a new opinion.

We must bear in mind that the formation of an opinion or the establishment of a dominant newspaper are processes that often take many years to come to fruition so that we no longer remember how preferred opinions or, more succinctly expressed, monopolies of opinion have originated. This applies equally to political systems, which are after all merely a certain manifestation of a public opinion molded in a political establishment. In extreme but unfortunately very real cases the community advances like a column marching arm in arm, step by step, deeper and deeper into a swamp. If an individual wants to drop out, his neighbors will not follow suit, and all will drown in the end. There is no doubt that from such an aspect the question of collective guilt appears in an entirely new light. Nobody really wanted the "final solution," but all stumbled into it nevertheless. We shall return to this question when we discuss the subject of dictatorships.

The Simplification of the World

Regarding the question of how opinions are formed as order parameters we have seen that in the field of natural sciences the order parameters appear very conspicuous and can be described in very few words, such as a "dominant laser wave" or the "honeycomb structure." With the aid of amplification processes nature succeeds in eventually establishing very clearcut states of order. We can trace the reasons for similar behavior during the formation of sharply outlined intellectual order structures with the aid of investigations by the American journalist Walter Lippmann. Primarily two factors encourage the occurrence of order parameters—that is, of uniform concepts. One is the limited resources—i.e., the limited number of news items or of trends that can be reported. This leads inevitably to a radical simplification of complex reality to a world of make-believe, which has been clearly expressed by Niklas Luhmann. In Walter Lippmann's words: "Every newspaper as it reaches the reader is the result of a whole sequence of selections."According to Lippmann this creates a suitable environment, we could call it a world of make-believe, for the reader; to put it even more concisely, what is not reported does not exist.

This process presents a simplified picture of reality, but one that we regard as the true reality. Here, then, is one thing that favors the occurrence of order parameters: the natural need for selection. The other reason must be sought in the fact that the selected subjects in the intellectual sphere can be as tersely expressed as the order parameters in the natural sciences. This is done with clichés or, as Lippmann termed it, with stereotypes, a concept familiar to newspaper printing press operators who cast the text in a rigid mold in the stereotyping room so that it can be printed as often as necessary. Stereotypes, then, are clichés coined deliberately to represent well-defined situations and often associated with a certain opinion, such as the cliché of "prohibited occupations" in the Federal Republic of Germany. This stereotype is the coin circulated and continously used, which helps a certain opinion to gain the upper hand over the competition. In Lippmann's words, "He who gets hold of the symbols currently dominating public feeling strongly controls the way to politics."

This contest between the order parameters, so widespread in the natural sciences, has not escaped the notice of the social scientists. Elisabeth Noelle-Neumann is only one of many examples: "Attention is brief, persons or subjects must prevail over strong competition. Pseudocrises and pseudonews are created by the mass media to eliminate the competition of other subjects."

What we are able to recognize with the methods of synergetics are the general laws on which all these processes are based. Opinion formation or publications are both subject to general laws that inescapably result in an enormous reduction of a wide range of opinions to a few. But the very knowledge of these laws enables us to take counteraction, as we have already pointed out.

Added to this must be yet another aspect, which is not reflected in the physical and chemical reactions but quite strongly in animate nature—i.e., the evolutionary character. We are faced with a continually changing environment. At the same time new ideas are continually conceived as others die. This is an enormously dynamic process that is also mirrored in the press. It will be enough to point out a few aspects from a sociological angle. The press may take up new subjects to start a process of public opinion, for example. According to Niklas Luhmann this requires the finding of words and of formulae. Eventually the subject becomes media-worthy, often with synergetic processes playing an important part. It is illuminated from the most varied angles or by vastly different newspapers. After a time the readers become saturated. Absolutely everything that could be said has been said about the subject, which eventually is, so to speak, beaten to death. But American communications scientists have made an interesting discovery; they investigated which precedes which, the views of the population or the subjects focused on by the press. They found that as a rule the subjects focused on by the media preceded the development of the nation's views.

In our outline of a few basic ideas of mutual influences we pointed out that the media are by no means absolute dictators but have their own problems of survival. There is a constant coming and going, manifesting itself in the publication of new periodicals and the foundation of publishing houses on the one side and in mergers and in the disappearance of periodicals on the other.

Television—Its Enormous Power of Influence

All this also shows that different mechanisms apply to television, particularly to state-run television, where there is no direct feedback from consumer attitudes to financial prosperity, let alone to the very survival of the service. On the other hand, television too is subject to the constraints of limited resources—i.e., a finite transmission time. Therefore, for example, it is usually impossible to broadcast the speeches of all the politicians in full. The necessary reduction—i.e., the selected passages of a speech—inevitably call for some pre-editing that may favor a certain opinion, whether or not this is intended by the television editor concerned. If we take the susceptibility of the viewer into account, interesting synergetic questions arise whose wide range has yet to be probed. If we assume, for instance, that the various opinions broadcast are represented according to their distribution among the population, according to the laws of synergetics it must generally be expected that a single opinion will be progressively reinforced, eventually gain the upper hand, and rule absolutely unless quite glaring contradictions in our environment give it the lie. If, on the other hand, all opinions of whatever shade are equally represented, the picture will also be distorted. Fringe groups are given undue weight and many of them will acquire undesirable support. The only way out of this dilemma may be to tolerate for a time the dominance of one opinion, to be replaced by another; but care must be taken to maintain freedom of expression and not to "freeze" an opinion.

Government and Public Opinion

Synergetics has shown that the order parameter has two aspects or a dual function. On the one hand it enslaves the subsystems, on the other it is upheld by them. We have seen that the order parameter that is "public opinion" is subject to this principle, but it has further functions—which was realized only gradually: popular opinion influences it, and public opinion also acts on the government. This is what David Hume (1711-1776) has to say on the subject:

> Nothing appears to those engaged in political philosophy more astonishing than the ease with which the *many* are governed by the *few,* and the readiness with which people subordinate their own feelings and wishes to the feelings and wishes of the government. When we try to analyze how such a miracle comes about we discover that the governing classes can find no other support than opinion. Government is founded on opinion alone; this applies both to the most despotic and most military and to the freest and most popular governments.

These considerations are summarized in one sentence of Hume's: "It is on opinion that government is founded." The clearest influence of public opinion on government is expressed by elections in democratic countries. Here an odd phenomenon occurs that at first seems to contradict what we have just said. In many countries elections more and more often end in deadlock. Approximately the same number of electors vote for the government or for the coalition supporting it as vote against. It is interesting to examine possible reasons for this because it is a phenomenon we have already met in synergetics—i.e., different answers to the same question or, expressed more clearly, various solutions to the same problem.

The behavior of the parties often resembles that of the ice-cream vendors on the beach described in the chapter about the economy. The parties also compete with each other, which sometimes may even put the survival of one party at risk, whereas the other party faces the question of gaining power or of retaining it. One should have thought that a party subscribes to certain ideals that it wants to put into practice on behalf of its voters. But the party very soon discovers that to reach its aims it must first gain power, which it can do only with a sufficient number of voters. It will therefore have to plan its election strategy so that it attracts voters from other parties. This corresponds to the behavior of the ice-cream vendor who moves his stand closer to that of his competitor until, at least outwardly, he occupies a position that the customer can no longer easily distinguish (at least before he buys his ice cream). Only after the customer has been obliged to eat the same ice cream four years running will he perhaps be able to notice the difference. This superficial assimilation, manifested often even by the same slogans—Peace, Freedom, Justice—being used by the different parties, reveals that criteria of decision can be established only with difficulty in a complex world. Furthermore, from an objective point of view, entirely equivalent yet completely different solutions may well exist for certain economic or social problems. Here one group will suffer badly while another benefits, and vice versa.

For these and probably also many other reasons the "synergetic input" or the synergetic curve we have come across repeatedly will have several locally preferred positions. One group regards one position as the better, the other group the other. This establishes the familiar symmetry. We know what is going to happen: very minute fluctuations or, in the political context, very minute groups or parties may tilt the balance and break the symmetry. If one configuration always retains the upper hand, this may result in progressive narrowing of the opinions. It is the foremost characteristic of democracy that at least in principle it offers the possibility of giving the other side a chance. To this extent democracy offers greater symmetry than dictatorship, in that it offers a much broader spectrum of opinion and of possibilities for individual development; in other words, democracy is able to guarantee a pluralistic society. This also ensures greater adaptability of democracy to changes in the environment, such as in economic conditions. All possibilities of reaction are latent, waiting only to be suitably reinforced to meet a new situation. Our observations on economic processes have shown that even here a new *state of equilibrium* cannot always be reached.

In general a democracy may be characterized by the fact that it is based on a fundamental order but leaves the structuring and self-organization to the individual or to the various groupings. This is not the order of a graveyard but one on a higher plane, which guarantees the freedom of the individual citizen and the variety of opinions this implies.

From the phenomenon of public opinion, its origin, its effect on the individual and on the government, we move our spotlight to dictatorship and examine how far one can speak of a public opinion in this form of government.

Dictatorships

Those who have lived under a dictatorship know that something like a dual climate of opinion exists. This phenomenon can also be observed in a democratic state like the Federal Republic of Germany, where it has been investigated and described by Elisabeth Noelle-Neumann.

In a dictatorship this "dual climate of opinion" is significant in that the government-controlled information services pronounce uniform opinions, reinforcing them with suitably selected news items. But there obviously persists beside it a, so to speak, personal but nevertheless somehow uniform opinion, which can therefore be described as public. This attitude, which strongly disagrees with the official line, is expressed in the whispered word and particularly by the political joke. The measures totalitarian regimes take against the spreading of these unofficial opinions, which nevertheless have the status of "public opinion," clearly show that personal contact between individuals can produce a climate of opinion that may threaten to become a danger to the government or to the ruling class. The regime's measures often include a ban on listening to foreign broadcasts or their jamming, restrictions on copying facilities such as the Xerox process, and strict registration of printing machinery and their output. This would suggest that only broadcast or printed opinion can put the regime in danger. But we know that under such a regime the secret police and a network of informers produce a general fear of expressing criticism in public. Any utterance of private opinions is thus prevented by intimidation. However, the eruption of collective attitudes cannot always be stopped. To cope with this possibility dictators install safety valves for public anger, often in the form of persecution of racial, religious, or dissident minorities that do not conform to the norm.

Through its investigation of collective effects, synergetics has also found the answer to why dictatorships are so stable, although this is totally incomprehensible to most citizens living in a democracy. The reason is the self-stabilizing effect of a large system. For its state of order to collapse, all or a very large section of the citizens would have to break out of the so-called state of order of the dictatorship *simultaneously*. But because dictatorships impose tight restrictions and controls on the communication facilities of the individual citizens, these can only make attempts to break out independently of each other, attempts that are doomed to failure because the other members of society happen at this very moment to support the old system or push in the most varied directions, thus hampering each other in their actions. The ground is prepared for a revolution by either a relaxation of government restrictions, which facilitates an exchange of opinions, or by the establishment of a secret underground network. But here, too, efficient dictatorships have taken countermeasures, for instance in the form of the agent provocateur who gains admission to an underground group as an active member, only to denounce the genuine members to the police.

Public manifestation of an opinion can nevertheless occur in a country like this, although this is apparently quite futile. Once, on board a plane in such a country, I noticed an interesting incident. The stewardess as usual was distributing newspapers. The passengers took them and immediately looked at the last page. I thought at first that it must contain particularly topical news until I discovered that it was the sports page. One could hardly imagine a clearer, yet in no way indictable rejection of the regime.

Public Opinion and Minorities

Synergetic laws in the most diverse fields of nature as well as of society generate selective pressures that result in an increasingly marked closing of ranks, for instance to form groups holding the same opinions. This tendency can lead all the way to the ostracism or persecution of dissidents, especially of those groups distinguished from the majority by external characteristics (such as race or religious adherence). These groups do not enjoy the protection of the state in every country. To survive, some members of the group react with assimilation by behaving indistinguishably from the majority, if possible. There is a second way for a group to behave for the sake of survival, an opposite way—by adopting a high profile. This spurs the minority group to progressively better performance; such better achievements will generally be held in high regard, as will the group, and its conditions of life will be safeguarded.

The social behavior of such a group, too will differ from that of the rest of the population. The problem of survival forces its members to support each other. The strength of the group rests entirely on cooperation. The attitude of the majority is mostly different. Here everybody sees a marked competitor in the other fellow. Rousseau's words in his prize essay "Discours sur l'origine de l'inégalité parmi les hommes" (Discourse on the Origin of Inequality of Men), which made him famous in 1755, apply to the competitor. He observes: "I would make it clear how much this all-powerful urge to acquire fame, honor, and distinction, which consumes us all, exercises talents and powers; and we can measure how much it arouses and multiplies passions, turns people into competitors, rivals, and, even worse, enemies."

Revolutions

Revolutions always interfere profoundly in the life of all citizens. From the aspect of synergetics the macrostate, the form of government, changes drastically. We have repeatedly described drastic changes during physical, chemical, and biological reactions. Time and again very clear analogies are evident in these processes. This allows us to expect that political or social revolutions can be investigated with the methods of synergetics. Accordingly, revolutions resemble phase transitions, such as from the nonmagnetic to the magnetic state of an iron magnet or from the disordered light of a lamp to the ordered light of the laser. But we must beware of rashly drawing parallels that are too close, for instance in the interpretation of the meaning of states of order in the sociological sphere; we would quickly meet with justified objections. We shall therefore use the term "state of order" only very loosely to characterize a form of government, with different forms of government corresponding to certain phases.

In inanimate nature, as in the liquid and solid phases of water, the mutual relations between the individual molecules are regulated in a certain way and in turn thereby determine the macrostates of ice and water. In the same way do the types of behavior of the members of human societies differ in different forms of government. Similarly to the transitions existing between various states of order in inanimate nature, for instance solid-liquid, we find revolutions of different kinds. They may change a monarchy into a democracy, like the French Revolution, or a democracy into a dictatorship, like Hitler's takeover of Germany; others lead from one form of dictatorship to another like the transition from the dictatorship of the czars to that of Stalin. Transition from a dictatorship to a democracy seems to be a rare event in this day and age, whereas that from one dictatorship to another is far more frequent. Obviously absent from this list is the transition from one democracy to another. In fact, it is precisely the property of a democracy to retain its basic character even when the party in power changes.

What are the mechanisms that are decisive in a revolution? We are now well able to define them synergetically with mathematical models and to identify them through the observations of historians. A revolution always seems to be preceded by destabilization, when the vast majority of citizens is no longer prepared to support the ruling system of government strongly, if at all. Added to this is the mutual susceptibility and mutual influence, which is perhaps the decisive factor. The negative behavior toward the ruling system thus becomes an avalanche, which can also be shown mathematically. This violent increase is further favored by an increasing self-isolation of the supporters of the ruling system and by their silence, thereby withdrawing their adherence from the system, partly deliberately, partly unintentionally. This situation is graphically characterized by Elisabeth Noelle-Neumann's term "the spiral of silence," a phenomenon that was observed and described before in, for instance, Alexis de Tocqueville's 1856 history of the French Revolution. De Tocqueville says of the decline of the French churches during the mid-eighteenth century:

> People who upheld the old faith feared that they were the only ones who had remained steadfast in it, and because they were afraid of isolation more than of the error they joined the masses without thinking like them. What was [still] the opinion of only part of the nation thus appeared to be the opinion of all, and precisely for this reason it applied irresistible pressure to those who gave it this spurious appearance.

What we have here is an amplification or reinforcement that is even more marked when most supporters have inwardly and spontaneously already deserted the system.

But what are the reasons for such a destabilization? It can be caused by economic misery due to a protracted war, by intellectual repression, by high unemployment, or by oppressive taxes (for instance the cause of the Peasants' Wars of 1525 in Germany), or by a vision of the future that is no longer realistic. In addition there are the already mentioned attempts of small groups to destabilize the political establishment by terrorist acts that create a climate of insecurity. Here, too, a spiral can be set up. When justice is undermined, for instance, the decisions of the judiciary are no longer backed by the state, and the prosecution of criminal acts inevitably becomes less and less convincing.

The theory of phase transitions has enabled us to rediscover in the mechanism of revolutions the effects found in the natural sciences. We know that the destabilization of the old state to be observed during a phase transition is connected with phenomena of strong fluctuations. The evaporation of water, for instance, is accompanied by violent density fluctuations. In sociology it is the accumulation of unusual events that occur rarely in the normal form of government and even then without any lasting effect. In the political arena this can be the rapid increase in terrorist acts, street battles between competing political groups, wildcat strikes with devastating consequences for the national economy, demonstrations, public meetings banned by the laws of the existing order. These manifestations are the signs of a crumbling establishment; often the way to the new order, to the new establishment, is by no means discernible, which can be proved by mathematical models. An example in recent history is the deposition of the Shah of Iran, when various political groupings competed with each other after his abdication. According to synergetics, a single highly active group can tilt the balance and determine the direction in which the whole nation will now be propelled. From the aspect of synergetics most revolutions are a symmetry-breaking instability. The mutual kindling of individuals' opinions, the urge toward revolution as a collective effect, can be observed most clearly during mass demonstrations. The mob rouses itself into a state of collective frenzy that creates an irresistible urge to do something, such as smashing shop windows, burning automobiles, or, as during the French Revolution, storming the Bastille. During such states of collective excitement individual logical thinking seems to be largely suspended. The individual looks as if enslaved by an order parameter—i.e., by an often haphazardly created slogan.

This leads us back to our initial question, whether revolutions can be predicted and even precalculated. Opinion research institutes already appear able, to a certain extent, to discover the existence of an excessive discrepancy between the opinions of the nation, between the actual and the ideal situation. It does not seem to be beyond the realm of possibility to forecast the breaking out of a revolution with the methods of synergetics on the one hand and through a refinement of demoscopic methods on the other.

We must, however, introduce a few basic limitations here, based on some highly important discoveries. The various examples of synergetics reveal time and again that the further development of a system is often no longer clearly predictable at its points of instability. Because even minor fluctuations may tip the scales, only predictions of probabilities can be made. It is also difficult, precisely owing to the necessity of triggering fluctuations to occur, to predict with any accuracy the time of such a revolution or popular uprising. This knowledge is obviously important also to those who want to bring about revolutions and in doing so to aim at a new dictatorship, and there is no doubt that these methods are also used by the super-powers when they interfere in the affairs of other nations. (1) The ruling political system, whether democratic or authoritarian, must be undermined. (2) A determined group of revolutionaries must be ready to push the nation in its destabilized state into the new direction.

Perhaps most readers will agree with these statements on the basis of their own thinking and observations. On the other hand it seems heresy to mention another consequence that can be derived from mathematical models, because it demonstrates clearly not only that the occurrence of certain macrophases (I want to avoid the term "state of order") is a macroproperty of the system, but also that the properties of the individual constituents can also be important. Expecting a laser to emit green light from atoms that without laser activity emit only red light will never succeed. Similarly we must ask ourselves whether the occurrence of certain forms of government is favored or discouraged by the character of the nation concerned. This is a vast field yet to be investigated by sociologists and social psychologists.

Can General Principles of Action Be Stated?

A few general conclusions, some of them disillusioning, can be drawn from the various examples of the natural sciences, sociology, and from mathematical treatment. Here are some of the most important ones.

Collective action alone, when one person does the same as his neighbor simply *because* the neighbor does it, may often produce the most diverse macrostates, forms of government in the political sphere, or modes of action that can include even collective crime. The murder of minorities may be as much part of the official policy ritual as the destruction of another species is for other kinds of life. A second aspect must be added to this. In economic and political decisions the solutions are rarely clear-cut; several equivalent solutions exist. Solution in this context does not mean that a new optimum state has been found. The disadvantages and advantages of one solution are equivalent to those of the other solutions. These alternative solutions are most intimately connected with the interlinked modes of collective attitudes. Because one individual acts in a certain way, his neighbor thinks he ought to react in another way. In the face of this situation we must ask ourselves how we can prevent the descent of the whole community to actions that the individual would condemn as criminal and how we can find clear-cut solutions to alternatives, which may not always be possible.

The only answer to this problem I can give personally is that we must base our judgment on higher aspects, that is, moral, humanitarian, or religious ones. This implies that expediency must be ruled out and that we must never wait for our neighbor to make the first positive move. If the individual is not prepared to make it himself he cannot expect his neighbor to make it.

Precisely because of the possibilities of alternative solutions, wrong turns that lead to catastrophe cannot be avoided with collective action, in which the individual is not guided by deeper insight. But every individual faces on his own such crossroads in his decisions. Is he to follow ethical points of view or is he merely part of a collective in the struggle for life and for survival?

Some Reflections on Bureaucracy

A phenomenon only just beginning to receive the attention of synergetic research is bureaucracy or, strictly, its continuous growth; with its ever-increasing labor costs it appears to be diametrically opposed to the functioning of economics with its recurring bouts of rationalization to improve efficiency. This calls for a focus on some of the causes of bureaucracy's growth.

We have pointed out in the chapter on the economy that profit is a decisive motive in the attitude of commerce and industry and directly linked with economic survival. This feedback of success is absent from very many administrative public bodies. Because these bodies produce nothing except reams and reams of paper, on the one hand it is difficult to measure administrative work with the yardstick of economic success; on the other hand, the administrative apparatus, especially of the government, is growing apace. This growth in turn causes major losses because of internal friction, if only because in a larger establishment more and more bosses' assistants become involved in a question, and the number of interrelations between individuals increases as the square of that of the staff. But this applies not only to government offices. The growth of administration in major companies is unmistakable and may sometimes considerably weaken their competitive position.

One aspect of the analysis of administrative processes soon shows that here basic principles of self-organization, which we meet again and again in nature, are completely neglected. A huge flow of information proceeds from management to lower levels and back, which from the point of view of the natural scientist must appear totally absurd. Whereas the actions at lower levels are increasingly regulated in the greatest detail, which involves an enormous effort to produce rules and regulations, even the most competent lawyer or administrative expert is no longer able to maintain a complete grasp of all the problems he is likely to meet. Thus regulations that are too inflexible become counterproductive and can result in inhuman decisions. It must, however, be admitted that too vague a formulation of rules may lead to arbitrary acts, for instance in the legal field, where some person goes free, whereas someone else goes to prison for the same offense. But we must examine the question whether many administrative procedures would not be much quicker as well as take much strain off human relations if a greater latitude of action were permitted.

The other, much greater expense of energy consists in the insistence on detailed control of every subordinate department, depriving it of all responsibility for its actions. This obviously multiplies the workload because the higher authorities once again, and sometimes several times, repeat the checks already conducted at a lower level. Control can thus be more expensive than the damage caused at a lower level by negligent practices and sometimes even by deliberate malpractice.

Numerous examples of synergetic systems demonstrate that control procedures in which higher authorities positively interfere in the activities at lower levels may produce chaotic situations—i.e., functional sequences whose actual effects are diametrically opposed to what had originally been intended. Those familiar with administrative procedures will be able to confirm this.

The answer from the synergetic angle is relatively simple. But it remains doubtful whether bureaucrats can be found who are prepared to accept it. Prototypes from both animate and inanimate nature quickly encourage us to permit far more self-organization at lower levels—i.e., to impose no more than general outline rules, which the various subordinate departments can supplement according to local conditions and on their own initiative. This can at the same time considerably reduce the flow of information. As nature demonstrates, transmission of all available information may not only be far from essential, it may also even be inconvenient; only relevant information is necessary. To take an example: the manager of a chemical plant does not have to know the details of the processes required for the manufacture of his products. Completely different parameters, such as the cost incurred, influence his decisions. It is the responsibility of his subordinates to introduce new production processes, and he cannot and must not interfere with their decisions. It is, after all, their specialty to search for the right method when new methods or new materials become available.

I readily admit my doubts that the growth of bureaucracy can be prevented unless the entire company or the structure of public administration collapses, when the whole cycle will start afresh.

Chapter 14

Do Hallucinations Prove Theories of Brain Function?

Probably the most complex and also most fascinating structure ever evolved by nature is the human brain. When a surgeon opens a human skull he finds a seemingly almost homogeneous gray mass, permeated by fine veins. In reality it is an unimaginably complex network of nerve cells.

In the second half of last century the Italian Camillo Golgi succeeded in making individual nerve cells visible by staining them. One among one hundred cells absorbs the stain and assumes the color of copper. We see a knot from which numerous branches radiate (fig. 14.1). A microscope is necessary to see these cells, also called neurons, which are only 1/1000mm in diameter. Their number in the brain is estimated at about 100 billion, the number of suns thought to make up our galaxy. They are held in position, supported, and nourished by other cells, called neuroglia. The neurons are often arranged in layers. Some researchers currently believe that columnar structures exist within and even between layers in which the cells are specially "wired up," thus forming functional units.

By wiring we mean the many connections between the neurons that permeate the brain like telephone cables or wires. Some go to neighboring cells, others like submarine cables go to far distant regions of the brain (fig. 14.2). These connections have indeed the same function as telephone cables. Like them, they transmit electric signals with a

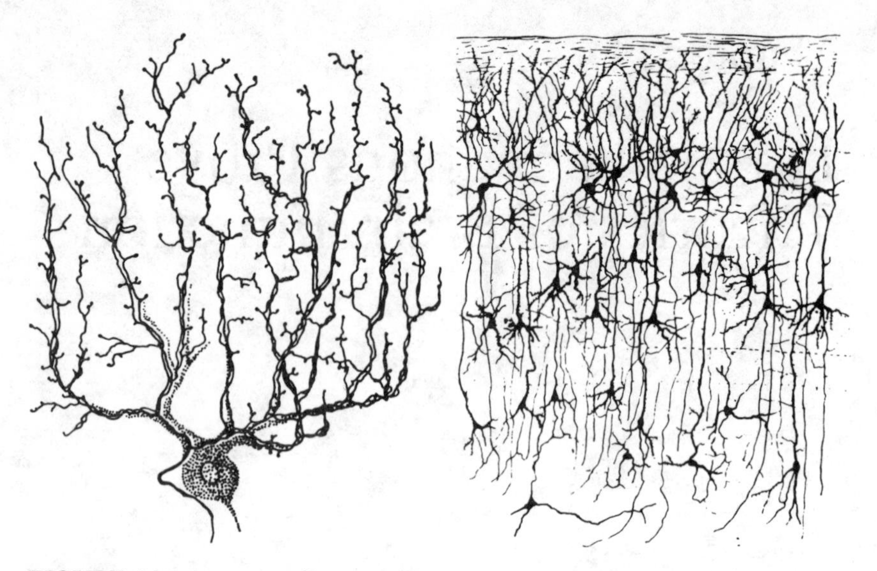

FIGURE 14.1 A nerve cell.
FIGURE 14.2 Network of nerve cells.

Morse instrument of their own. But whereas human Morse code consists of dots and dashes, that of the brain is composed of dots only. Nature, however, has found a way of transmitting information of different content: it transmits the dots at different frequencies. The neurons are capable of processing the information as it arrives and of passing it on to other neurons.

Fine electric wires (electrodes) introduced into the neurons from outside enable the neurologist to scan the electric events inside a single nerve cell.

Does the "Grandmother Cell" Exist?

Science has certainly a long way to go before it can unravel human thought processes. But some interesting experiments have been devised that shed light on the functioning of at least a few cells or regions of the brain. H. Hubel and T. N. Wiesel conducted experiments with chimpanzees. They showed them objects (such as bands or bars) or moved them in front of the chimp's eyes. The eye of the chimpanzee accepts the light impressions and transmits them to a certain region of the brain that is responsible for vision. When the researchers inserted their electrodes in this region and investigated the reactions of the individual nerve cells to certain objects placed in front of the eye, they made a surprising discovery: certain distinct cells react to certain external impressions. Some cells, for instance, respond not only to a bar, but also at different strengths depending on the orientation of the bar; i.e., a cell transmits a large number of Morse dots, or, in the expert's jargon, discharges many nerve impulses when the bar has a certain orientation. When it is turned through about 90°, the cell practically ceases to react (fig. 14.3). Cells were found that respond in a certain way even to the movement of bars. It appears that these nerve cells belong to a higher level of the brain, and that the items of information arriving from the eyes are prepared and processed by series-connected nerve cells to produce the special reactions discovered in the investigated neurons.

In other words, it looks as if calculating processes are carried out whose end product appears in the cells investigated. Expressed in plain language, the statement reads "the bar is vertical" or "this bar is horizontal."

These findings could tend to reinforce an earlier hypothesis about the functioning of the brain, the theory about how the brain manages to recognize patterns. This postulates special cells in the brain in which not only a bar but, for instance, an entire face can be recognized as

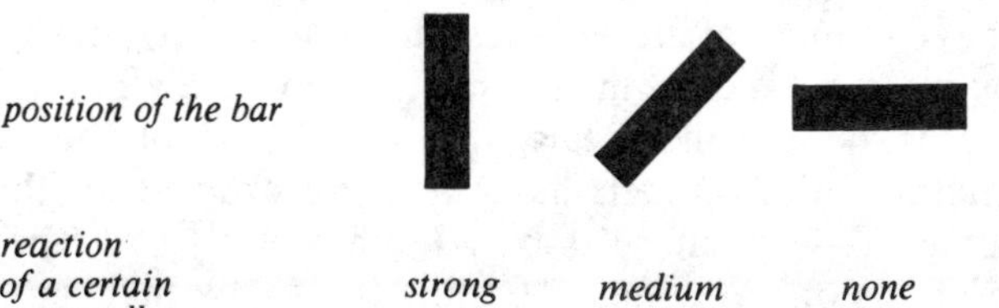

FIGURE 14.3 Reaction of certain nerve cells to the position of a bar in the field of view.

such. These hypothetical cells are jocularly called "grandmother cells" in the literature. They are supposed to enable a person to recognize his or her grandmother, for example. The majority of scientists, however, have abandoned this hypothesis. An intensive search did not reveal cells even in the chimpanzee—which does recognize patterns composed of, say, bars. The investigation of brain damage, caused by an accident, let us say, has also shown that thinking and memory processes are not strictly localized in the brain but distributed across large regions. Science today tends to assume that the perception, memory, and thinking processes are typically collective effects in which always a large number of neurons participate. But when many neurons perform functions collectively, we must obviously ask ourselves how this can be proved.

Before we deal with this question, we must eliminate a possible misunderstanding. From what we have said one might conclude that the various abilities—e.g., hearing, sight, or even language—are distributed throughout the brain. This is by no means so. It has been known for a long time, from accident research if nowhere else, that strictly defined regions of the brain are responsible for certain functions such as vision, hearing, smell, and also for speech. The speech center, by the way, consists not of a single block but of two blocks, one of which appears to be responsible for the form or grammar, the other mainly for the content of the speech. New medical aids have made it possible to render the function of the various parts of the brain visible. If a part works more actively, it requires an increased blood supply, which can be labeled with chemical agents (space forbids discussion in detail) to make any increase in perfusion visible, often with an x-ray camera (although the physical reactions here are different). This has revealed that different human activities switch on certain blocks of well-defined zones of the brain (fig. 14.4). It is a profoundly interesting event in which a large number of individual systems act synergetically.

But the question we want to examine is what happens in an individual part of the brain, the one that deals with visual perception, for instance. A few mathematical models have been constructed for the functioning of such parts—e.g., those assumed to have only two types of neuron—on the basis of experimental results that show that some neurons reinforce nerve impulses and that others have a kind of inhibiting function, in other words they suppress signals. At first thought the effect of the inhibitory neurons appears puzzling. But they have an important function in that they enable us to recognize such things as blurred outlines as edges. We do not wish to bore you by going into details here. We are instead interested in the question of how such brain models can be basically tested.

Excitation Patterns in the Brain—Hypotheses and Experiments

We have shown, especially in the chapters on physical and chemical processes, that the most varied systems can produce identical patterns. We see both in liquids and in air the same rolling motions of molecules, for example, which arrange themselves macroscopically. We have found time and again that the various interrelations between the constituents of the system are not all that important. The same macropatterns occur whenever we reach the points of the system's instability.

When the American biomathematician Jack Cowan heard of these analogies during one of my symposia on synergetics, especially of the formation of rolls in liquids, he had a bold idea. He linked hallucinations with the formation of macropatterns of excitation in the brain. Persons who had taken drugs such as LSD reported very similar perceptions. They saw, for instance, concentric circles or outward-moving rays or spirals (fig. 14.5). Cowan had developed a mathematical theory earlier about how the image recorded by the retina is transmitted to a flat layer of the brain that is responsible for visual perception. Such images can be explained as follows: individual nerve cells situated in the retina, called receptors, receive the incident light and transform it into nerve signals. (Let us ignore the question of whether this is done by an individual cell or by a whole complex of cells.) This point in the retina transmits the signals to a definite point in the brain through a nerve fiber. Neighboring points in the retina have their own special "telephone wires" to the brain and they too reach neighboring points in the brain. But if we apply Cowan's postulate—that the circular field of the retina is imaged on the quadrangular field of the brain—we make a surprising discovery: the images seen during hallucinations correspond to straight bands in the brain—i.e., the excitation patterns of neurons, differing only in their orientation (fig. 14.5). Cowan even succeeded in tracing still more complex perceptions during hallucinations back to their original patterns in the brain, to the already familiar honeycombs.

How can all this be explained? The taking of drugs obviously desta-

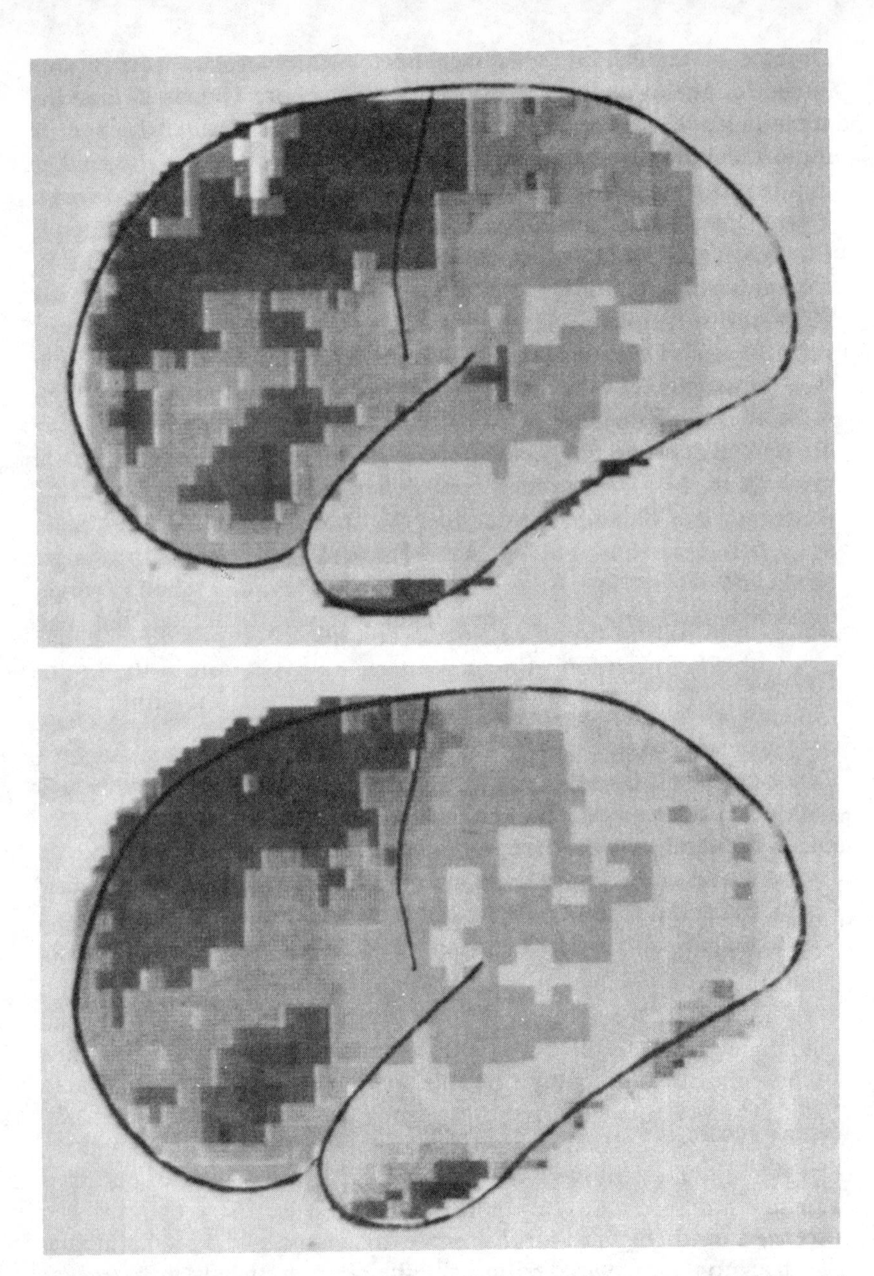

FIGURE 14.4 The block-type switch-on of the brain in various activities, such as motion and speech, is clearly apparent from the level of blood supply going to individual parts of the brain.

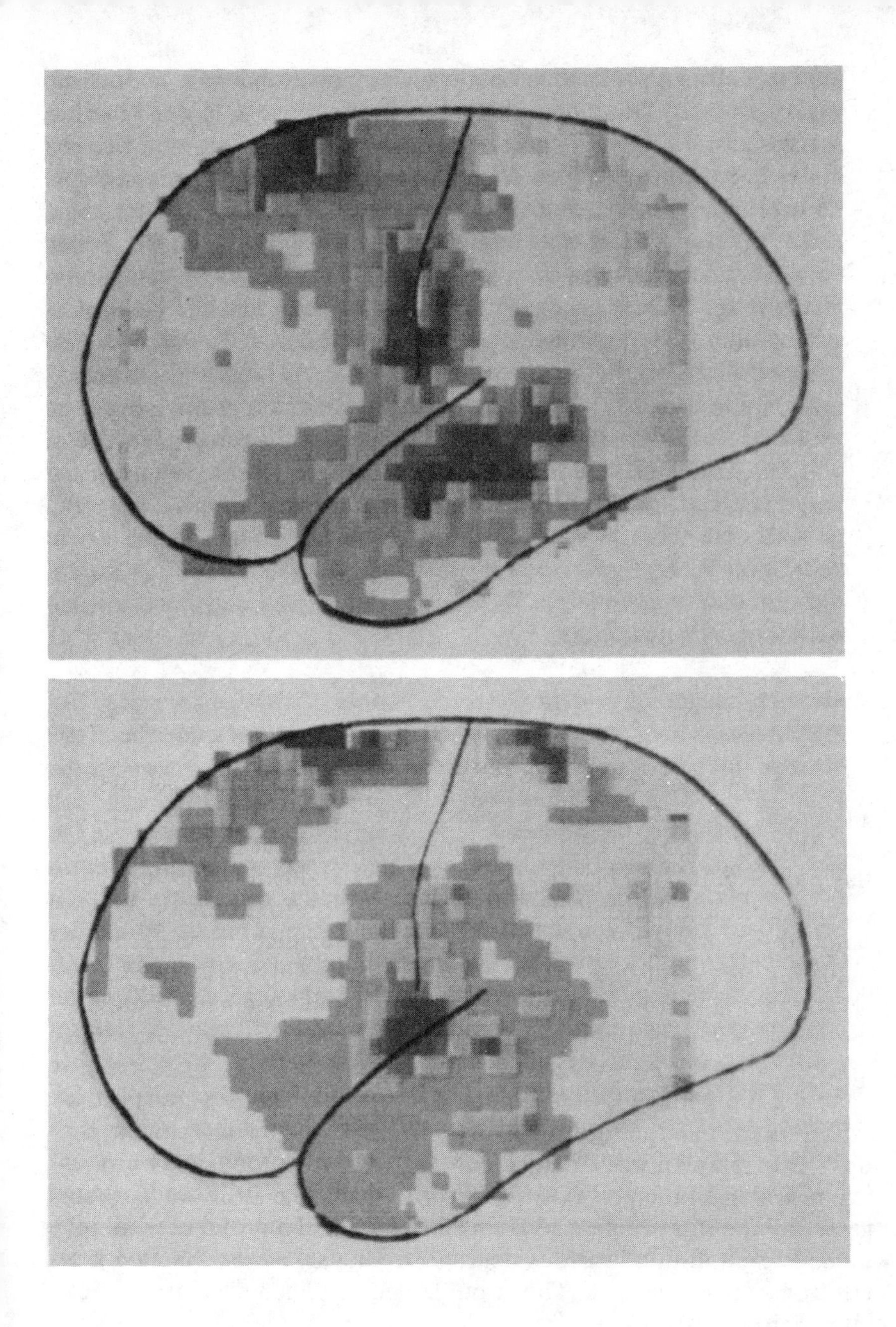

FIGURE 14.5 Cowan's theory of hallucinations. *Left:* Example of perception patterns of persons after taking drugs. *Right:* The bandlike excitation patterns in the brain according to Cowan's hypothesis.

bilizes the function of the brain; the original state of rest is pushed into a position where it is to give way to a new macrostate, here an excitation pattern. This situation closely resembles that of liquids that initially are at rest but when heated from below suddenly assume a state of macromotion. Beyond the drug concentration that destabilizes the brain, the neurons start to fire like mad. Interestingly, though, not in mad *confusion,* but in a totally *orderly* fashion. Obviously we do not claim that the brain moves physically like a liquid; we have tried only to offer a lucid explanation of an analogy exclusively based on mathematics.

At the moment these ideas must still be regarded as pure speculation. It is quite in the cards that they will be proved by experiment, but obviously this requires an advance from present methods of neurological research. In the past only the firing of a single neuron was investigated, by the introduction of a single electrode. To prove such simultaneous excitation of various neurons, obviously a whole network of electrons must be used. This would seem to open fascinating new approaches to neurology.

Our assumption—that many neurons simultaneously fire their

impulses in well-defined patterns—may be thought highly hypothetical. But a phenomenon exists in neurology in which the correlated firing of many neurons is indeed observed. The firing is so uniform as to generate an electric wave that is measured with the encephalograph. These regular brain waves occur during epileptic seizures (fig. 14.6). Regular excitation patterns (here periodic oscillation) are thus linked to a pathological process. From this aspect, periodic oscillations during epileptic seizures are analogous to the hitherto hypothetical spatial patterns in the brain during hallucinations. It is interesting to note that the coordination of many nerve cells indicates pathological behavior. But we must not conclude from this that thinking is unconnected with correlation effects. Quite the contrary, without doubt. If we imagine the neurons as tiny lamps that light up when excited, we would continuously see the most varied lamps light up and go out, and it would be like solving a jigsaw puzzle to find out how the lighting up of the individual lamps adds up to an overall picture that makes sense, similar to the picture on a television screen. For the time being we must still use relatively indirect indications of this simultaneous action of many neurons.

Thinking in Blocks

Much speaks in favor of the assumption that thinking occurs in blocks. When we learn a foreign language through living in a foreign country

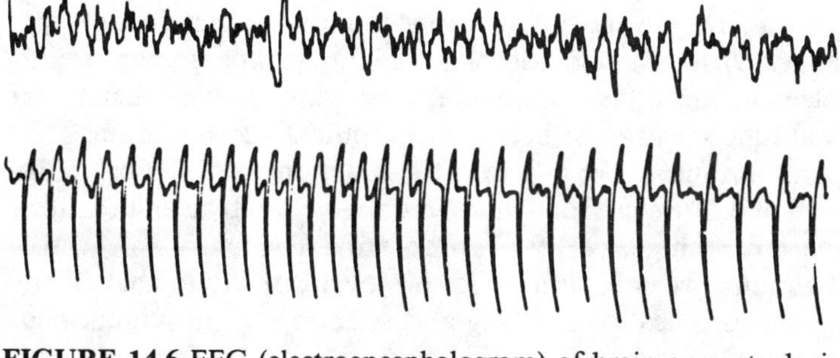

FIGURE 14.6 EEG (electroencephalogram) of brain currents during normal brain activity (*top*) and during an epileptic seizure (*bottom*).

we find that we often pick up not only whole sentences in the form of turns of phrase but also separate words. Modifications and exchange of individual words will then enable us to form new sentences with new contents. Nevertheless we do not want to recommend the exclusive individual-word or the exclusive whole-sentence method, because when we learn how to spell the very opposite effect is decisive—i.e., analysis, where each word appears from the very beginning dismembered into its components. However, this is only incidental.

Thinking in whole blocks is also familiar to chess masters. Sixteen white and sixteen black chessmen, some of which have different functions (bishop, pawn, knight, king, queen, rook) face each other on the chessboard. The beginner learns the various possible moves of the pieces and during the game contemplates the various moves and considers the consequences, whether he can defend his rook or mate his opponent's queen. Chess masters, however, think in entire sequences. They see the pieces in their mutual configuration, which tells them their next moves without their having to trace the moves of the individual pieces in detail. Conversely, a chess master may find it very difficult to think in terms of individual pieces if some wholly new situation forces him to do so. There is an important difference in this type of thinking in blocks between a chess master and a chess computer. You can now buy chess computers against which you can play and even select the degree of difficulty of the game you want to play; they are for sale even in department stores. Today the best chess computers can be beaten only by the best chess masters, such as grand masters. This might suggest that these machines are far more intelligent than human brains. But the way they achieve their objective of beating their opponent is surprisingly primitive. They simply try out all the possibilities in a large number of steps and then compute how they can most effectively decimate the number of opposing pieces that still carry a certain weight. Obviously a plodding procedure compared with thinking in whole configurations. This example provides clear evidence that fundamental differences exist between computers and the brain.

Another important aspect bears on the functioning of the brain. By putting more and more neurons into a circuit we produce an organization of progresively higher level of complexity. The brain is nevertheless able to return from such a *collective* level directly to that of the single cells, by specifically actuating and selecting an individual neuron, for instance.

Synergetics makes creative achievements appear in a new light. As in a jigsaw puzzle, an entirely new integrated image forms in front of our eye. Something like a phase transition of consciousness occurs in our brain, much that previously was disjointed suddenly appears as something rationally arranged, and vexing reflections suddenly give way to the relief of certainty. The new insight had long been latent in our consciousness, and suddenly it strikes us like a flash of light. We cannot escape the impression that here processes are at work similar to those we already know from other fields of synergetics. A fluctuation ("the flash of light") creates a new control (i.e., the new idea) that subsequently succeeds in subordinating and correlating the various aspects; in enslaving them. But all this happens in a completely self-organized manner—our ideas, too, organize themselves into new insights, new knowledge.

It may be this very fact that explains so many processes of self-organization in nature.

Body and Mind

The synergetic approach also opens up a new way of dealing with the problem of body and mind. To discuss this question let us start from the opinion of the famous neurologist Sir John Eccles (1903–), which he expressed at the 1980 Lindau meeting of Nobel Prize winners. He sees a possibility of solving the body-mind problem by treating all parts of the body as merely interchangeable accessories. The essence of a human being could therefore be reduced to certain regions of the brain. According to Sir John, the ego is the programmer and the brain the computer, that is merely the executive organ. Synergetics has reached a different conclusion. Its concept, which is represented in this book, is that the order parameter and the enslaved constituents or "subsystems" are conditional upon each other. According to this interpretation, the order parameters are our thoughts, and the subsystems the electrochemical reactions in the neuron network in the brain. We have seen many examples of the interdependence of the functions and existence of the order parameters and subsystems. This is how synergetics views the relation between body and mind in the last resort.

One last word on the interpretation of brain function through models of the brain: apparently the latest field of science to develop is always used analogously to describe the functioning of the brain. In the past it was electric circuits and even the mechanical train of gears; we talked of the "switchbox of thoughts." Today, needless to say, we use the computer analogy. What will it be tomorrow?

Does the Brain Grow According to a Plan?

Because of the extraordinary difficulty of obtaining information about the complex reactions in the brain, researchers tried to find another way: they investigated the question of how the brain grows. Is there a definite predesigned plan? Briefly, this is the present state of knowledge: during the development of the embryo the so-called neural tube, which consists of cells, is formed first. The neurons grow in the vicinity of this tube; they are produced, as it were, as on a conveyor belt in a factory. But they do not remain in this place; they are moved to other regions of the brain, where they diffuse into their new locations in a way quite similar to that of the individual cells of slime molds. According to an American embryologist, they collect in layers and first form something resembling an ant heap. But how do the individual cells know where they have to go? Very little has yet been discovered about this. But there are indications that the individual neurons travel along the formed neurologia to reach their final destination.

Another development strongly reminiscent of the behavior of the slime mold exists. As in the cells of this mold, an attractant entices the individual neurons. It is the so-called neural growth factor, which is produced in particular regions and diffuses through the tissue. When certain neurons notice this substance they migrate in the direction of its source. During these migrations some cells may well lose their way and sometimes die. And occasionally it happens that the brain may become diseased because of wrongly incorporated cells; but we are interested only in the further growth of the healthy brain. Ramifications grow from the individual neurons. Some search out other cells in the vicinity and establish contact. Others grow far toward other cells of the developing brain. Without any doubt the development of the neural network in the brain is self-organized. As far as we know the connections form automatically without any higher control. Researchers disagree about the operation of this self-organization, but the various opinions may well all be correct, although they apply to different parts of the brain or to different species. Let us compare just two views.

One view claims that the growing ramifications can, with the aid of particular molecules, recognize the cells with which they have to establish connections. This could be likened to the individual neurons having locks which can be opened only with certain keys (i.e., with the individual excrescences). It is often found that initially more "telephone wires" were formed than were eventually used. These will be retracted, or they die off like some individual neurons that have not fitted themselves properly into the network. In this concept the eventual network is a well-defined circuit, which proceeds according to a preconceived plan laid down in the molecular locks and keys.

The other concept is more favorable to the ideas of self-organization. It claims that the links between the cells form in great confusion. But when nerve impulses from the sense organs reach this network the development of certain links will be favored over the development of others, depending on the degree of their use, automatically; the network thus grows, along with its functional efficiency, only during and through use. The idea that use—e.g., the processing of perceptions—reinforces connections in the nervous sytem is known in the literature as Hebb's Synapse. Synapses are junctions installed between the nerve cells, like substations. These are the components that are said to be reinforced through frequent use. Unfortunately there is as yet no direct experimental evidence that often-used synapses are larger than others. The very idea—that a nerve network forms only in the course of its being used—greatly fascinates the designers of computers. Is it possible to build computers that organize themselves largely automatically during their activities? We shall return to this question in the next chapter.

Chapter 15

The Emancipation of the Computer: Hope or Nightmare?

The Wunderkind *of the Twentieth Century*

It is a current trend to replace human labor increasingly with machines. Not so very long ago the basic idea was to free people from the drudgery of menial work or at least to lighten their burden. The many household appliances such as washing machines, dishwashers, and vacuum cleaners come to mind. Similarly, machines have taken over in the factory, where monotonous work like wrapping bars of chocolate has long been taken over by machines. But astonishingly, research and development have recently turned more and more to the task of also replacing mental activities with machines. Computer technology is a typical example, although the computer's ability to "think" genuine thoughts is often grossly overestimated. All computer programers know that a computer is stupid and cannot eradicate even the simplest error by itself unless it has been expressly programed for it. Employees in offices converted to computer accounting know a thing or two about that. Suddenly the entire entry has disappeared without a trace in the computer and simply cannot be found. Nobody subsequently can teach the computer how to come up with the required entry unless it has previously been instructed in the greatest detail to do so. Computer intelligence, however, is a question we shall deal with later in this chapter.

But apart from this or that shortcoming computers perform astonishing feats. The computer is perhaps the most revolutionary item of twentieth century technology. Whereas in the past we heard about its scientific uses, especially in space travel, we now meet it everywhere. Whether we reserve a seat on a train or book a flight, look for a prediction of election results or even for a partner, we use the services of a computer. We find it not only in offices but also increasingly in the home, where even our children will no longer want to be without it. Where in the past we needed a logarithm table or had to add long columns of figures, all we have to do now is press a few buttons. Built into an automobile, it is designed to save gasoline; in telecommunications it ensures the optimum utilization of cables. It plans the construction of buildings, tells us where doors and electric sockets should be, and draws the buildings in the most varied perspectives, even with trees in their surroundings. Computers calculate bridge constructions and design housing developments as well as chemical refineries. In flight simulators they are used for the training of pilots and astronauts; they guide rockets to the moon and to the outermost planets of our solar system. They control not only machine tools but also entire complex production processes; it is perhaps a futuristic aim of centrally controlled economic systems that here, too, a super-computer may plan and control everything. But this is the very point where the limitations that the experts call "information bottleneck" make themselves felt. A very simple example will explain this.

In most apartments room temperature is automatically regulated by a thermostat set at a certain temperature. A sensor continuously measures the room temperature, the so-called actual temperature. If the actual and set values disagree, the thermostat signals to the central heating boiler to raise or lower the temperature of the hot water. If we apply the whole principle to a manufacturing process, let alone to an entire economic system, the following basic problem will arise: a very large number of actual values must be measured, and the computer must calculate the controls to be executed for the rated values to be obtained. But this calls for a very difficult and time-consuming calculation that may take so long that the computer cannot give the control instructions in time; the whole control will collapse because the information is unable to flow through the bottleneck quickly enough. The answer to this problem in some cases may be supplied by even faster computers, but the general solution must be seen in the self-organization of part processes, when only certain relevant magnitudes must be input to let the operation proceed organically and as a whole. We shall also meet the problem of self-organization with the computer. To recognize the potentialities and the possible limitations of the computer, we must investigate it a little: how does it work "in principle," how do we use it for our ends—in other words, how do we program it? Let us start with programming.

Programming

Basically a large computer works in a way that differs little from that of a small pocket calculator. If we want to add 3 and 5 on the pocket calculator, we must press button 3, the plus button, button 5, and the sign of equation button to tell the calculator that it must produce the result. This means in more general terms that we give the computer the following instructions: take a number (5) another number (3) and add the two up; present the result.

The entire procedure can be divided into two large groups—into the selection of the special numbers (5 or 3) and into the calculation itself. We can treat the selected numbers like spheres bearing these numbers and housed in individual drawers. The process of calculation can thus be expressed as follows: take the number 5 from drawer 1 and add the number 3 from drawer 2 to it. Transfer the result to another drawer, say drawer 3. For a difficult calculation you can continue this process by multiplying the number from drawer 3 by one from drawer 4, and so on. The basic task is thus always very simple, but acquires a great latitude of variation because we vary the numbers from the drawers 1 and 2 or, alternatively, are able to continue the process. We may end up by saying to the computer: return the result you have obtained to drawer 1. It is thus possible to incorporate so-called loops, when the computing process is continuously repeated. In this way, for instance $2 \times 2 \times 2 \times 2$ can be calculated.

The programmer has to perform these two different tasks with scientific computers as well. He must state what individual computing steps (such as addition, subtraction, multiplication, division) the computer has to take, and provide for a store of numbers to enable the computer continuously to execute the same sequence of computations but always with new data.

The real work of the programmer therefore consists in the setting up of the computing steps. Feeding the computer the individual data, however, is on the whole easy. All you have to do is place the individual numerical values in the various boxes. Often a short program consisting of relatively few computing steps is sufficient, but many data can be processed, such as the calculation of bank interest, insurance contributions, wages, and many other items. With more difficult computing problems—i.e., longer programs—the need for highly qualified operators may become quite enormous in certain circumstances. This raises the question of whether manpower can be saved by the construction of self-programming computers. But we have not quite finished with human programming yet.

Although the individual computing steps are extremely simple, they can be combined in the most varied ways. At the same time the already mentioned loops—which are designed for example to repeat an approximate calculation until the result appears accurate enough—can be built in. Such calculations are, for instance, the extraction of the root of a number, an example mathematically minded readers might be interested in.

But this is not all a computer can do, and here the hopes and also the difficulties begin. As we store numbers in the various boxes to have them ready for further processing in the computer, so we can store signals that represent a certain computing rule. Here, then, the computer is instructed to draw the next ball from box 3 and to do what the ball indicates (obviously the computer does not draw balls as in a lottery). It has a data storage from which it calls up from certain positions electric signals that present it with instructions: "Do this [or that]." One "ball" can mean "multiply the two numbers input," for instance. It can also mean "execute [a particular very complex, computer] program." This linkage of various instructions may make computer programs highly complex, occasionally producing a certain variety of the species of programmer known as a "chopper." A chopper is a programmer who, like a do-it-yourselfer, continually devises new programs, finally loses control, and sits, red-faced, in front of his computer until late into the night to wangle ever new computing tricks out of it; but he gets more and more confused and in the gray light of dawn finds that total chaos rules in the computer. This shows clearly and without doubt that computer programming is full of nasty pitfalls.

The difficulties encountered by the choppers are by no means unknown to ordinary programmers. A leading computer manufacturer therefore had the question of large computer design investigated from an architectural angle; but these investigations have had no results, which is not surprising from the synergetic point of view. After all, the computer not only contains rigid structures, but events also continuously take place in it that are matched with each other and must interact. In other words, a computer is a typical synergetic system.

Computer Networks

Today, instructions are fed into a central computer from terminals in many separate rooms and the results displayed on the screen or printed out by the printers in the terminals.

But a new tendency is gaining more and more ground; it promises to open up an important field of applications of synergetics. Instead of a single, large central computer, many small interlinked computers are to be built to take over the functions of the earlier large computer, which like a master controlled the small; these are now to "converse" with each other (fig. 15.1). The advantages of a system consisting of many small computers are obvious. Small computers are easily mass-

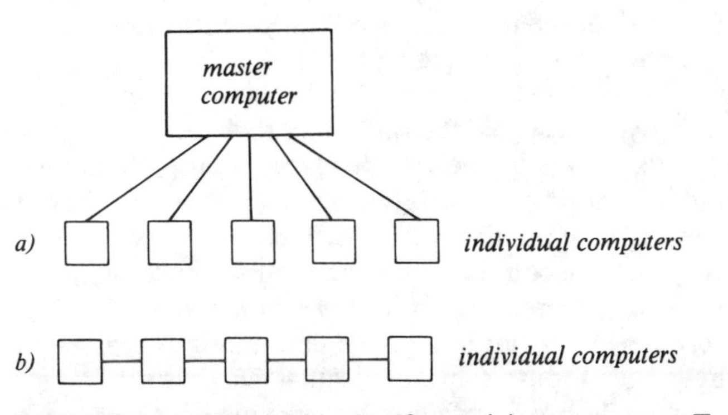

FIGURE 15.1 Organized and self-organizing computers. *Top:* A master computer allocates the tasks: organization. *Bottom:* The individual computers allocate the tasks among themselves: self-organization.

produced and interchangeable, therefore easier to maintain, and can be distributed in the various rooms of a computer center or of other establishments. Some of these computers may be identical, others may be equipped for special tasks, one with a monitor screen, another with a printer, and so on. This raises new basic problems connected with the question of how much self-organization these networks of small computers can be expected to cope with. On the one hand they could be prewired, which would combine them into a large computer again. Here the allocation of functions among the individual computers is determined by the wiring.

There is another possibility: that the various computers spontaneously establish new links among themselves. Thus a computer that needs assistance would have to send signals to another computer. Such signals carry flags, as the computer people say. They contain the information about the source and destination of the message and also "ask" the other computer if it is ready to accept the task. The addressed computer then has to inform the transmitting one whether it is ready to accept the task. After a few interchanges the task is transferred. This obviously involves some complications that are unnecessary with prewiring.

It is the aim of the computer designers to find the best possible compromise between the rigid programmed linkage and the self-organizing distribution of functions, although the latter objective is still in its infancy. Thus computers are to distribute their functions and compute in parallel to arrive at new distributions of functions, and so on. Computer experts here speak of the installation of "deep-seated structures" intended to bring about such self-organization processes.

What are these "deep-seated structures"? The results described in this book may already help us a little, although they are not "deep-seated structures." As we know, we can produce macropatterns, for example in liquids or during chemical reactions, when we change external conditions, e.g., supply the system with more energy. We can expect exactly the same with computer networks. By setting more tasks for the computer, such as increasing the number of the input data, we can expect entirely new distributions of the computing processes to occur in the individual computers automatically—i.e., in a self-organized manner. This may, however, produce undesirable phenomena we are also familiar with in synergetics, such as oscillations. In this situation the distribution of functions between the computers fluctuates periodically, which causes an enormous flow of data between the individual computers. Here, too, the computer expert can consult synergetics to learn how to eliminate such oscillations. Furthermore, he will find analogies with the neuron network of the brain useful. Frequently recurring function distributions can thus be employed to let initially loose connections develop into permanent ones. Here the principles of competition, of the "survival" of the most efficient link and the suppression of all the others, come into play. This may lead to the computers eventually beginning to think in blocks like good chess players (cf. our previous chapter). These blocks are not necessarily confined to a highly specialized computer; they may include several computers.

Finally one could think of imposing conditions on the whole computer network that act like Darwin's principle of survival, for instance that a given problem has to be solved in different ways until the network uses only the quickest solution, let us say.

We appear to be fairly close to the realization of these ideas, provided we change the problems set only a little at a time. As in other synergetic systems, new "structures" may suddenly be produced with the result of a novel distribution of functions in the computers.

The situation becomes difficult when the computer network has to solve an entirely new problem. No miracles should be expected here. Like human beings, the network must first test various possible solutions, learning all the while.

Since computers can do so many things we must ask ourselves where the computer, compared with man, is weakest. Certainly not in the identical processing of many data. The stumbling block is . . .

Pattern Recognition

Pattern recognition is essential to many automated processes. An automatic welding machine, for instance, must find out where a certain piece must be welded. The task becomes still more interesting when the machine has to recognize even more complex patterns. A familiar example is the reading machine, which is able to decipher and to recognize written texts. Here, too, synergetic effects play a decisive role. As a first step let us imagine a letter split up into certain components, so-called primitives or elementary properties (fig. 15.2). These primitive elements have been chosen so that they can be perceived by the machine, for instance as a bar or an arc, situated in a certain place and curved in a certain way. Such primitives can be "scanned" with the aid of photoelectric cells, and "recognized" by means of relatively simple circuits. A number can be allotted to each of these primitive elements in a certain position (fig. 15.3). As only a certain number combination will open a combination lock, so a certain number combination that corresponds to all the primitives of a letter will determine this letter. The machine must therefore check whether it has a number combination in its register which represents, for instance, the letter A. The problem of this recognition arises from the possibility of errors in the whole system, perhaps that one of the primitive elements could not be positively identified and a vertical stroke was confused with, say, an arc open to the right. This takes us back to the old problem of how to correct an incorrect sentence.

Such procedures are familiar to us from the laser and from liquids, where initially it was quite possible for some subsystems to break ranks. Some of the laser atoms were able at first to emit "wrong" waves, for instance, or not all the molecules of a liquid participated in the rolling motion. But the order parameter very quickly incorporated them in the general order. A machine that has been given wrong information, which is of course not contained in its list, must look for the correct information that comes closest to the wrong kind. Individual mathematical methods can be developed for this purpose. They are very lucid and consist in the allocation of points to the numerical details as in a coordinate system (fig. 15.4) and the measurement of the distance between such points. Here, too, by the way, the system of symmetry break can occur (fig. 15.5). A wrong combination can be equidistant from two correct ones. Here the machine will be helpless, unless it is allowed to use new criteria of decision. When a written word or sentence is to be read by a machine, which cannot decide whether a letter is, say, O or X, we can make the decision, as we know from human activities, only by looking at the word or even at the sentence as a whole. Grammatical rules, and sometimes only the meaning deduced from the context, allow a decision about the correct nature of the letter. This example shows that character recognition from one to the next step can become an unusually complex process at the very moment when contents become decisive.

The method discussed here is relatively rigid, because the primitives considered, for instance the individual arcs of a written text, must be

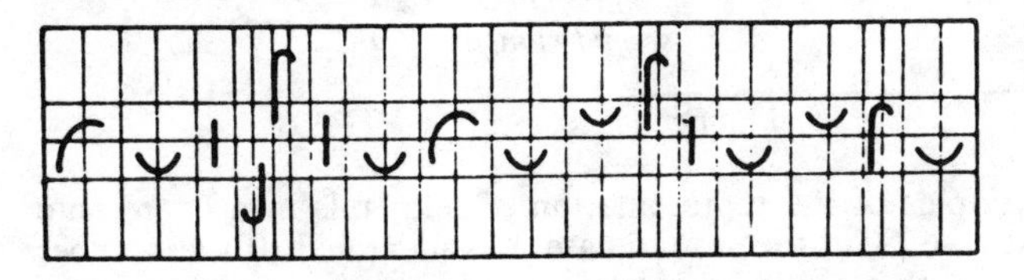

FIGURE 15.2 Pattern recognition through division into simple components (primitives).

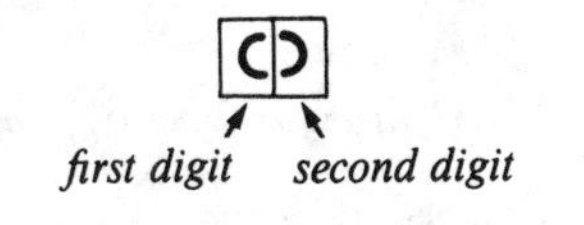

number combination: recognizes letters

ↄ → *1*
ɔ → *2*

12 ⟶ *O*

first digit *second digit*

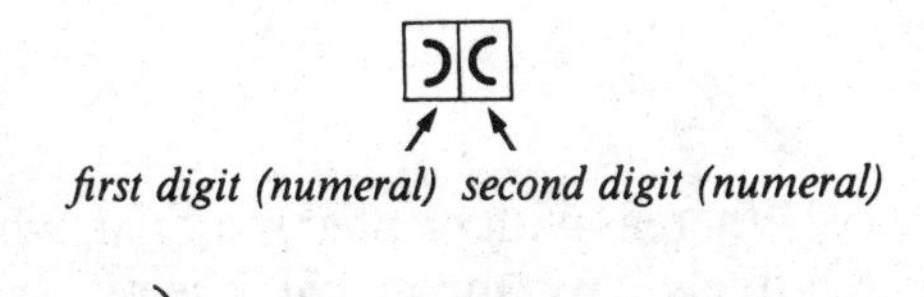

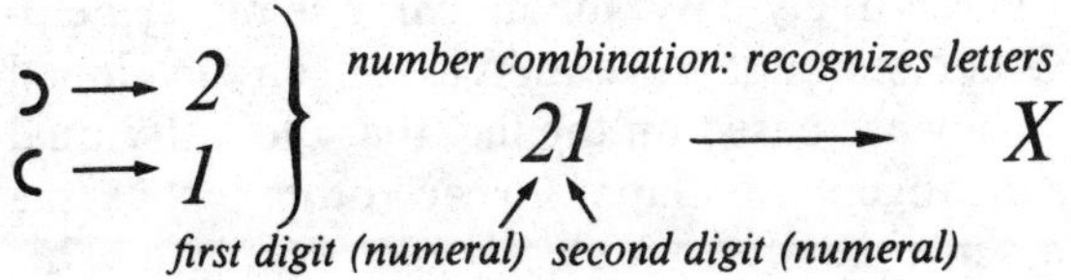

FIGURE 15.3 Simple example of the correlation of numerals with primitives in certain digits.

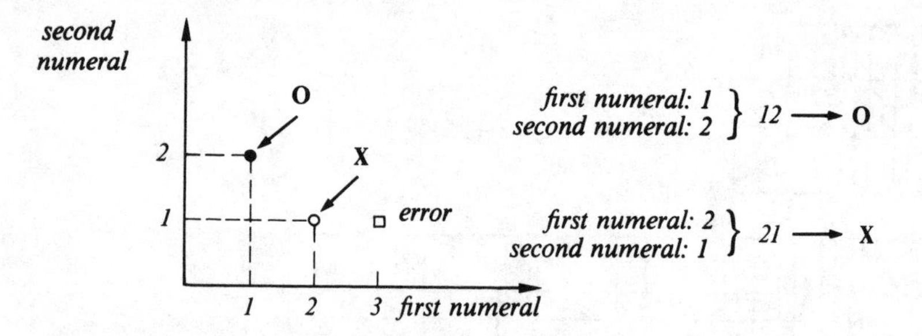

FIGURE 15.4 Examples of the representation of numerals in a coordinate system. In the first case *O*, in the second case *X* is identified. If the number combination were 3, 1, this would be an erroneous letter. In practice we must work with a high-dimensional coordinate system.

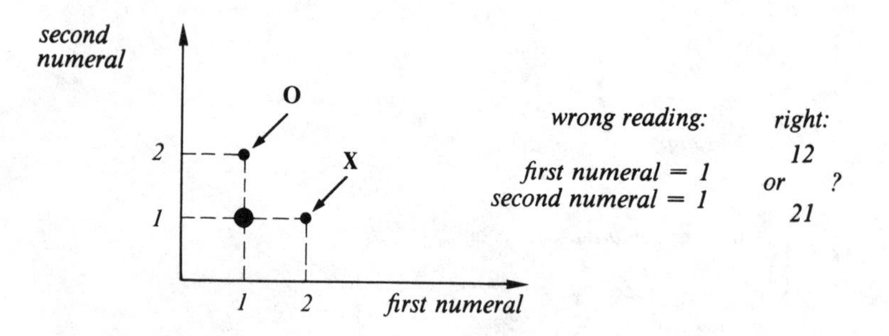

FIGURE 15.5 Example of an error. The machine has read 1 at the first and second digit, but the only correct letters would be *O* and *X* respectively as in Figure 15.4. The erroneous letter is equidistant from the two correct ones. A symmetry break is needed for the final decision.

read in a rigidly defined place. Whereas with printed material, which is fairly standardized, this is relatively easy, the machine is rather helpless with handwriting. Other methods have therefore been developed to recognize characters. They are based on the fact that the individual arcs must be in a certain arrangement relative to each other just as in a language the words of a sentence are in a closely defined place. This analogy between the grammer of a language and the arrangement of the primitives is indeed used to establish rules for the machine, which it must follow when it assembles the primitives. Time and again it faces a bifurcation as in a maze, but the arrangement of the various primitives points the way to putting the letter together.

Perception

What is the meaning of perception in this context? We allot to a letter that we perceive as A a certain number in the computer. But this number can also be used for other instructions. A letter-sorting machine, for instance, allots to the word "New York" a certain number combination according to the various letters it consists of. This combination can now be evaluated in the computer to instruct the sorting machine to "send the letter to the conveyor belt for New York" (fig. 15.6).

NEW YORK → *number combination* → *control instruction*

FIGURE 15.6

A very interesting feature of human perception is our ability to understand garbled information; that is, the brain can independently supply missing pieces of information. This appears to be an important component of our powers of perception. Even forms that are only hinted at are automatically compared with forms that are familiar to us and completed accordingly; here again, the interaction between individual items of information plays a decisive role. It has become possible within the last few years to construct machines that are capable of doing this, in a way similar to the recognition of a letter. The machine compares a given incomplete picture, for instance by dismantling it into primitives, with a picture available in the form of a reference, which it also dismantles into primitives. It will then look for the reference picture that most closely resembles the given picture. Having found it, it reproduces the entire picture, thus restoring the incomplete one (fig. 15.7). This ability of machines is also called associative memory. The decisive factor is the linkage of the various elements into a whole, which can then be suitably completed.

But machines ought to recognize not only the written but also the spoken language. To enable them to do this we convert the spoken

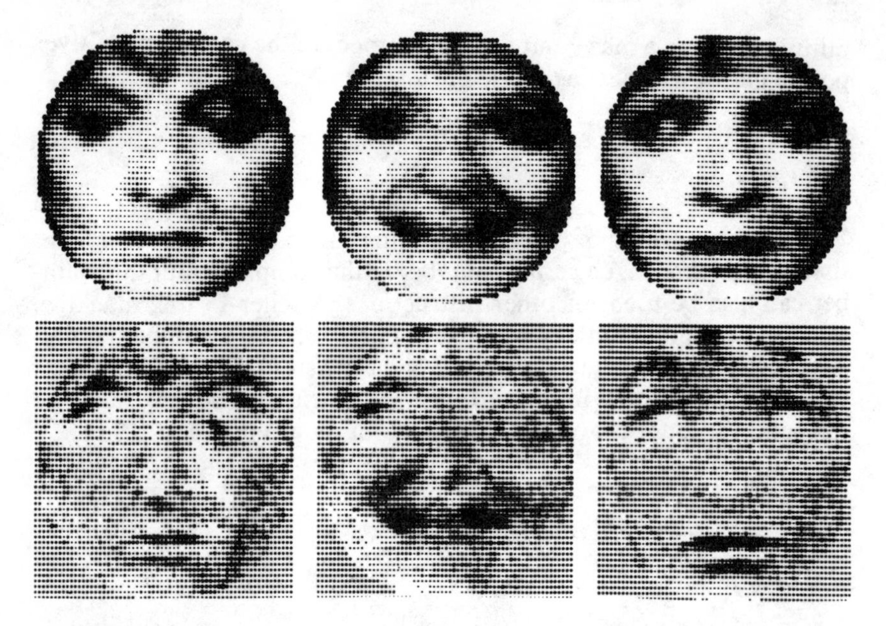

FIGURE 15.7 The reference patterns in the top row, representing various facial expressions of the same person, are stored in the computer. The bottom patterns are presented to the machine, which will then be in a position to reproduce the patterns in the top row.

sounds into electric oscillations and make them visible on a television screen. This produces a graph in which a perfectly characteristic sequence of peaks corresponds to a certain vowel (fig. 15.8). The machine compares these characteristic forms with reference patterns, so that eventually a spoken letter A can be transformed into a written letter A. It has, by the way, recently become possible to realize also the reverse process of transforming letters input as on a typewriter into sounds. As all these examples show, the elements identified by a machine are generally translated into certain number combinations. This clearly recalls processes in the nervous system where the most varied sense impressions are converted into nervous impulses—i.e., into equivalent electric signals. The nervous impulses are therefore the universal code used by the nervous system. Consequently we must not be surprised that perception processes in computers are nothing but signals in the form of electric pulses which transmit number combinations, for instance.

As we have already mentioned, the next step, from the identification of a letter or of a word to the association of this word with a meaning, is qualitatively new and enormously difficult to accomplish. We become aware of this particularly when we want to use the computer for the translation of a language. We have seen that the computer can convert every word into a number combination, which it uses to look in a kind of dictionary for another such combination, which corresponds to the word in the foreign language. The computer employs this second combination to print out the word of the foreign language.

But the real difficulty arises when we have to deal with the finer points of a language—i.e., when several words of a foreign (target) language correspond to a single word of the source langauge. We are faced with such a situation in a primitive way when we encounter a word

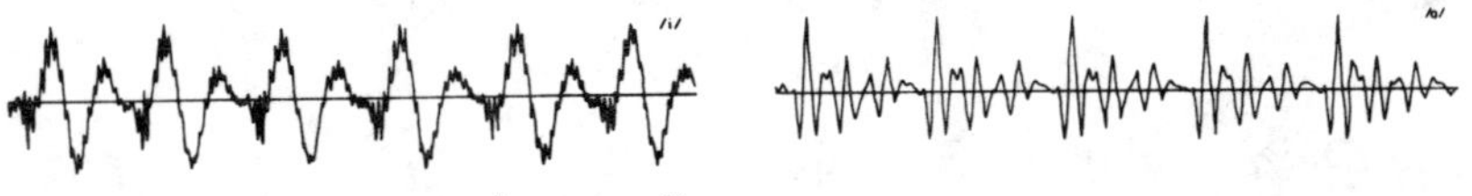

FIGURE 15.8 Acoustic wave pattern of several Anglo-American vowels (the time is entered horizontally; the amplitude, vertically).

that has several meanings: a match may be a contest between baseball teams, something to light a fire with, or mean harmony between two people. Here we face again the fundamental problems of breaking a symmetry: we have entirely equivalent meanings to choose from. The correct one can only be determined from the context. But how can a machine establish a context? This ambiguity makes it very clear indeed that here we are confronted by a hierarchy of problems that involves us in questions of ever-increasing complexity. From the aspect of synergetics the machine faces the task of finding a suitable hierarchy of order parameters. The individual words must thus combine to produce a meaning, they establish an order parameter, which is often also capable even of "repairing" (within certain limits) garbled sentences in the same way as the laser wave is capable of calling to order an atom that breaks ranks. Sometimes several order parameters have to be associated with a (written) sentence if it has several meanings. Here the machine must move a further step up in the hierarchy to determine the order parameters unequivocally. The difficulty often consists in the need for the backup of an enormous fund of human experience at the higher level if a text is to be "correctly" interpreted.

The Underworld of the Computer

Up to now we have dealt mainly with that part of computer technology known as software. Let us now examine how the computer works in detail. The part that interests us now is called hardware by the experts.

The computer divides the computing steps, or what is also known as logical thinking, into tiny steps. Such simple steps could be "and," "or," "yes," "no," or the storage in a memory—i.e., in a set of tiny boxes.

Such logical functions can be carried out even by quite simple devices. In public parks we often see mobiles powered by water (fig. 15.9). Water runs into a bowl, which tips as soon as it contains a certain volume of water. The water is distributed among other bowls, and so on. Initially we find the movement of the bowls quite irregular, but after some observation we realize that each bowl tips according to strict rules, which ensure logical steps. A simple example of such a mobile consists of two vessels that can be filled with water, and that we connect as shown in figure 15.10 to enable the water to flow into an overflow vessel. From there it ultimately reaches a fourth, last vessel. If the two upper vessels are empty, the lower vessel will also remain empty. If only one of the upper vessels is empty, the lower vessel too will remain empty. It will be filled only when both upper vessels are full. We can also formulate this as follows: provided the upper vessels 1 *and* 2 are full the lower vessel will also be filled (fig. 15.11). This is perhaps one of the simplest demonstrations of the logic link "and." Both preconditions must be met.

Many processes in life follow this link. If we boil an egg, 1, the water must be boiling, and 2, the egg must remain in the boiling water for a minimum time—e.g., 3 minutes—to be done. (This example is not quite

FIGURE 15.9 Mobile.

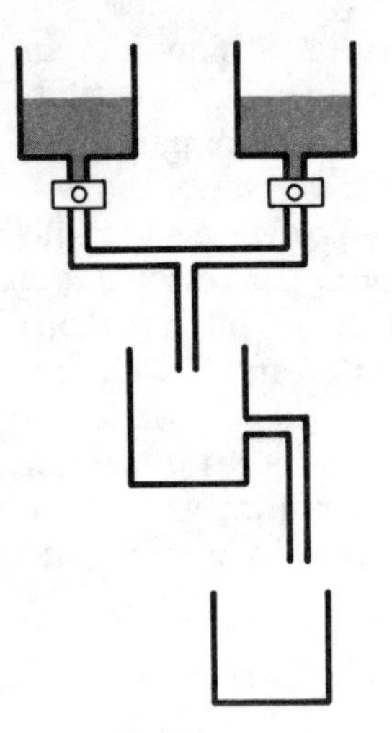

FIGURE 15.10 Example of a circuit that realizes the logic process "and." The bottom vessel will be filled only if both top vessels were originally full.

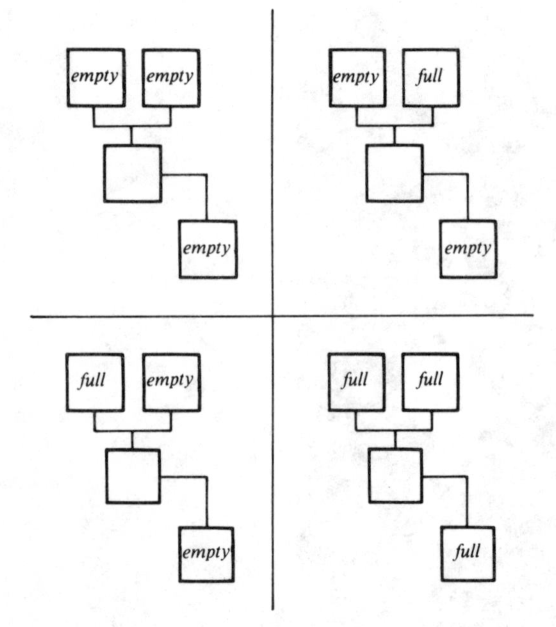

FIGURE 15.11 How the vessels manage to realize the logic linkage "and" (i.e., "not only _____, but also _____").

perfect in that individual preference decides when the egg is done. This is not so in mathematics, where such things observe a strict logic.)

Another mobile is a beautiful example of the "or" relation. It is practically the same device as the one just described, except that the outlet into the bottom vessel is at the bottom of the middle one. An overflow ensures that surplus water pours away. Here the lower vessel will fill up even when only one of the upper vessels has been filled (see figs. 15.12, 15.13).

Mathematical logic shows that all logic links can be represented with very few of such elementary steps as "and," "or," "yes," "no." But we need not dwell too long on these abstract ideas; let us see how such logic links can be used for practical calculations, that is, numerical ones. To do this we must descend into the underworld of the computer, which to the layman is a magic world.

The mathematicians show that all numbers can be expressed in noughts and ones; this is the so-called binary system. The calculating rules for all numbers can also be reproduced in this system as those of addition, subtraction, and multiplication.

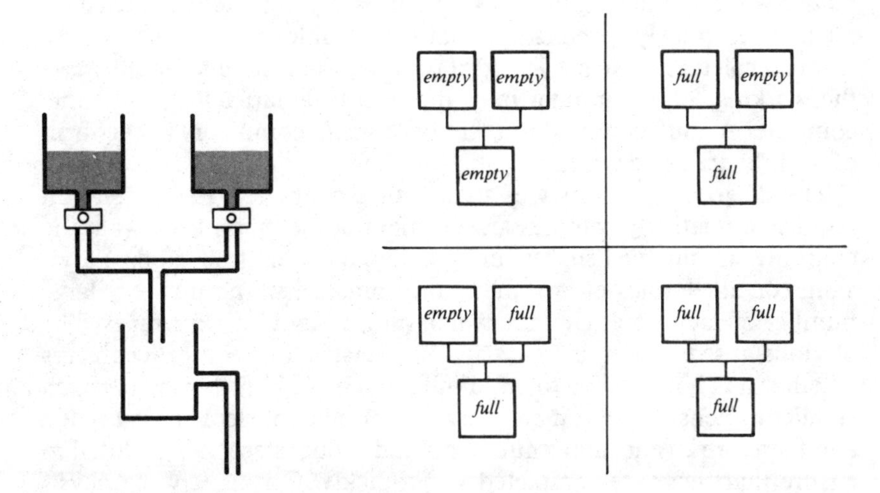

FIGURE 15.12 Realization of the "or" relation. Here the bottom vessel will be filled if only one of the top vessels was full.
FIGURE 15.13 The various possibilities of Figure 15.12.

We now find to our surprise that the individual parts of a computer communicate with each other in a kind of primitive language. The computer signals consist of only two figures, 0 and 1. The logic link "and" enables us to see at once how a computer can multiply these numbers 0 and 1 by each other. We want to watch the computer as it multiplies and to convince ourselves that it does it correctly. For this purpose we must check whether it produces the same results as we would otherwise obtain. As every schoolchild knows, $0 \times 0 = 0$, $0 \times 1 = 0$, $1 \times 0 = 0$, and $1 \times 1 = 1$. These rules can be precisely copied by our water computer, where empty vessel = 0, full vessel = 1. If we leave the two upper vessels empty, the lower vessel, which indicates the final result, also remains empty. We have thus verified that the computer behaves according to rule $0 \times 0 = 0$. If we fill one vessel with water, leaving the other one empty, the lower vessel will also remain empty, we find rule $0 \times 1 = 0$ confirmed. If we fill both vessels, the lower vessel will also become full; this proves the rule $1 \times 1 = 1$. We have now already grasped the multiplication table, as it were, of the computer. Other rules of arithmetic, such as addition, can also be realized with such water mobiles.

Here the water circuit is a little more complicated. The interested reader will find the diagram (figs. 15.14a and b) instructive. Such simple devices quickly demonstrate that they enable the computer to perform all calculating steps. The idea of using water mobiles to illustrate the workings of a computer may appear a little farfetched. But some computer manufacturers do in fact offer small computers in the form of such hydraulic devices.

The diagram also shows us that even the simple "and" operation requires a relatively complex arrangement of the pipes. If we want the computer to put even slightly more complicated multiplications, divisions, or other rules of arithmetic into practice we need a very large number of such water circuits, which may easily fill a skyscraper. The obvious answer, then, is to ask the physicists and electrical engineers whether it is not possible to build other circuits of this type with much smaller dimensions. But if an enormous number of elements is needed and therefore an enormous number of individual steps, it is essential to ensure that these are completed very quickly. Fortunately the physicists have long known how to build circuits not only of water but also of other elements. At the beginning of this book we mentioned electrons, those smallest particles, that carry the electric current through

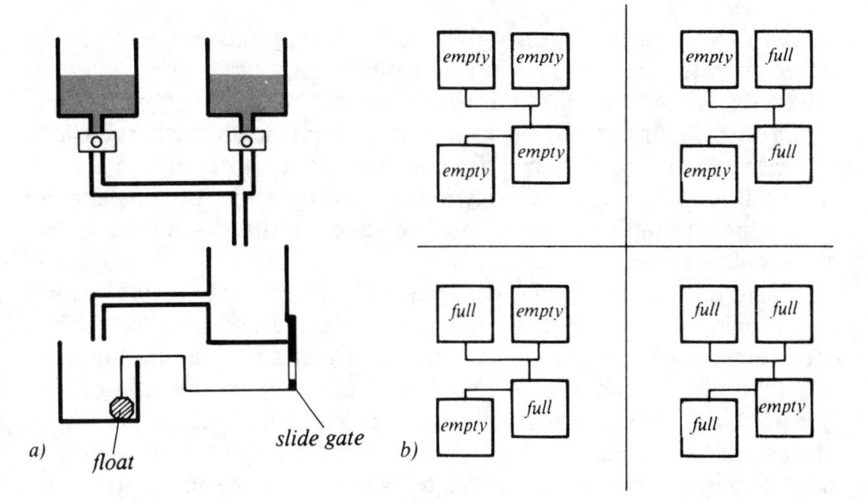

FIGURE 15.14 A water circuit for addition.

Figure 15.4(a) shows how the mathematical process of addition can be realized. A full top vessel = 1, an empty top vessel = 0. Filling of the bottom vessels represents the numbers in the binary system. From now on we shall speak only of the bottom vessels. If both vessels are empty = 0. If the vessel on the left is empty, that on the right will be full = 1. If the vessel on the left is full, that on the right will be empty = 10 in the binary system. In the conventional decimal system, this number = 2.

Figure 15.14(b) shows how the rules of addition are realized with the aid of the arrangement of Figure 15.14(a). If both top vessels are empty = 0 x 0. Obviously the two bottom vessels will also be empty. If the top left vessel is empty but the top right one is full and the faucets are opened below both of them, the bottom right vessel will be filled, while the bottom left one will remain empty. This represents the result of the addition = 1. Needless to say the same consideration applies if the top left vessel is full and the other one is empty.

Of particular interest is the case when both the top vessels are full; i.e., when we want to add 1 to 1. If we empty only one top vessel, the bottom right vessel begins by filling up. If we add the contents of the second top vessel, the bottom right vessel will overflow, filling the bottom left vessel. The float now pushes the slide gate up. This empties the bottom right vessel, and we obtain the end result shown in Figure 15.14(b) at bottom right, which corresponds to the representation of 2 in the binary system. This circuit is indeed capable of performing an addition in the binary system. Suitable combinations of such circuits enable us to add even more complex numbers than 0 and 1, but the basic principle is always largely preserved.

metals. These particles not only can be transported but also can be stored in batteries and capacitors, familiar to all of us.

Exactly as the water flows from one vessel into the next by the force of gravity, electrons can be transported from one vessel, such as a capacitor, to the next if they are allowed to "coast down" an electric potential gradient. This analogy between water and electrons enables the engineer to build all the circuits we have just discussed with the aid of electronics.

Within the last few years technology has performed veritable miracles in building ever smaller circuits. In the sixties, computers still had complements of radio valves, in appearance and size resembling electric light bulbs, each of which could execute only one switch function (similar to the structural element of our water mobile). With 18,000 valves the American ENIAC computer weighed 18 tons and cost almost a million dollars to build. Today tens of thousands of structural elements, each executing the function of the earlier radio valves, are assembled on wafer-thin sheets of 1 cm diameter, each costing ten dollars. Simultaneously switching times have been continually shortened. At the moment they are of the order of

$$\frac{1}{100{,}000{,}000} \text{ sec.}$$

Semiconductor elements, the so-called transistors, have replaced the valves not only in radio and television sets but also in computer technology. With the aid of superconducting circuits, the so-called Josephson junctions, a new generation of computers often no larger than cigar boxes is now being built; but to be operational they have to be stored in low-temperature compartments near absolute zero. Even if their temperature rises by only a few degrees they lose their memory and their capacity to "think," which is many times that of the current computers.

Logical Processes—Independent of the Substance

This is where synergetics again comes into its own, with its concept of the order parameter. Synergetics shows that these order parameters are themselves subject to the processes of logic. In the cases mentioned here, for instance, the densities of the electrons in the various parts of a structural element are order parameters characterizing the macrostate. With these circuits we have enabled them to interact with each other and to set up new ones. The interesting feature of synergetics' results consists in the possibility of performing such switch actions between order parameters in the most varied ways, often by a system on its own, without this circuit having to be installed at all. Thus computer circuits can now also be produced with laser light, promising switching times of the order of a trillionth of a second—i.e., 1/10,000 of the already unimaginably short switching times of today.

Computer circuits can also be based on chemical reactions. These considerations are of particular importance as we look for computer elements even smaller than the current ones. We thereby arrive at the dimensions of atoms and molecules, where animate nature reveals the prototype of the smallest imaginable computer element, the membranes of the cells, especially of the nerve cells. These membranes consist of elongated molecules, of which each has a kind of head and tail, with one part reaching out in the direction of a water surround, the other turning away from it. Their common aversion to water, then, forces the molecules into an alignment like soldiers in rank and file (fig. 15.15). Such membranes can also be produced artifically; they are only as thick as a molecule is long. Certain other molecules have the effect of opening or closing pores in such a membrane and thus permit the controlled passage of electrically charged atoms or other molecules (fig. 15.16). This produces switch elements that can perform logic funtions. It may not be long before we succeed in building computer elements, perhaps even whole computers, of these atomic dimensions.

The control processes on this microplane conform to the laws of synergetics. They are almost exclusively effects possible only through the cooperation of many individual parts. This opens fascinating vistas of logic processes that include thought processes able to run their course on quite diverse substrates, which may be water, electrons, chemical reactions, lasers, or biomolecules. These considerations take us down into the underworld of the computer, which we have "x-rayed" for its individual elements. But we must not make the mistake of thinking that all thought processes must be connected with such small logic elements. Perhaps there are other possibilities of the function of thought processes of which we have as yet no idea, for instance those that do not lend themselves to being divided into tiny elementary steps.

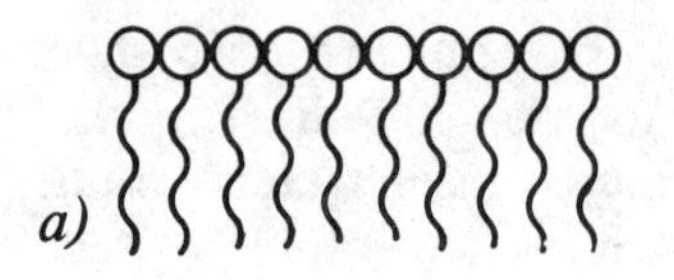

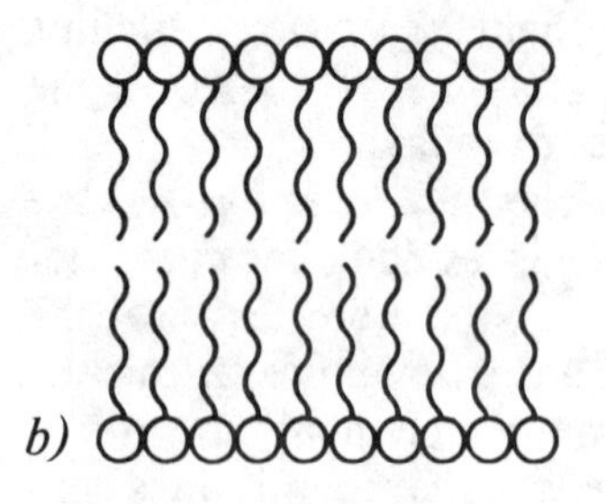

FIGURE 15.15 Examples of membranes: (a) of a single layer of molecules; (b) of a double layer of molecules.

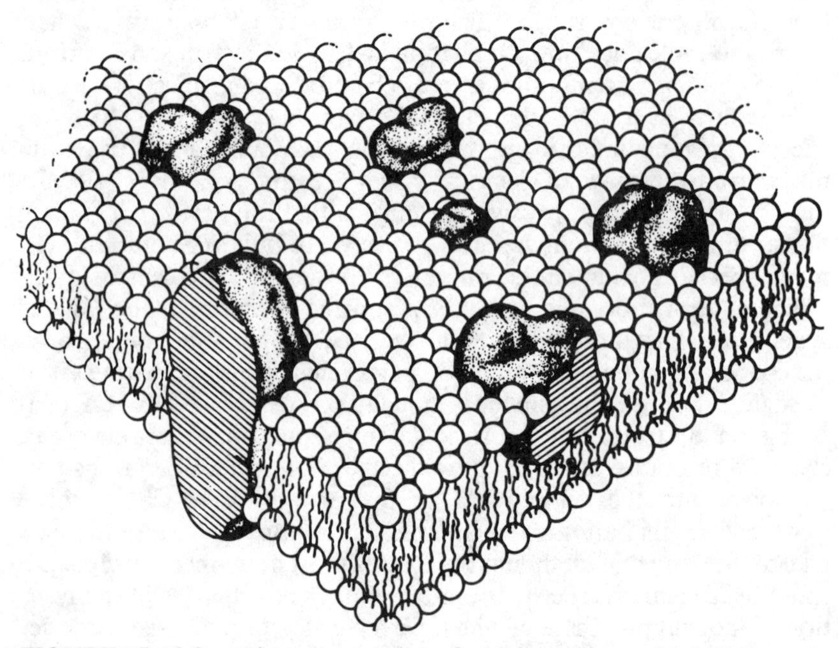

FIGURE 15.16 Graphic representation of a biomembrane with built-in molecules.

Can Computers Be Capricious?

We look on computers as inanimate machines that proceed in a strictly programmed manner, leaving no freedom, no uncertainty. But in this book we have already encountered problems that have no unequivocal solution, for instance when symmetry has to be broken, as with the ball in the bowl with two depressions—where does the ball go? In the perception—vase or faces; which of the two features do we perceive? If the computer is presented with such tasks, it will be frustrated, and we must call in coincidence to solve the problem. We must nudge the computer with a few random fluctuations to continue computing or toward further problem solutions. But even if we have not deliberately programmed these fluctuations, the solutions offered by the computer may look capricious. This must be expected especially with complex processes that correspond to those in psychology—where, after all, conflict situations succeed each other all the time.

This prompts the question of how far a computer can execute thought processes on a higher plane. The example of the recognition of sentences already demonstrates that the degree of difficulty increases enormously even if we take only a single step up in the hierarchy of letter—word—meaning of a sentence. Here the ability of the computer to establish lateral connections or, in other words, associations seems still far inferior to that of the brain. But the possibility cannot be ruled out that one day here, too, the computer will overtake the brain not only in the quantity of the material it can process but also in the manner of processing it. For the time being, however, it is still a pipedream, the existence of a branch of modern research called "artificial intelligence" notwithstanding. But let us give our imagination free rein. Some computers can already speak to a certain extent, others are able to recognize simple spoken words or sentences within limits. How far, then, can a computer come to resemble man? Can it have feelings? Does it have consciousness? These questions are outside the strict precincts of synergetics, which is mainly a science-oriented discipline. We will nevertheless offer a few ideas, not so much to provide the reader with ready-made answers as to stimulate thought.

Why can we human beings speak of feelings at all? On the one hand because we experience them ourselves, and on the other because we can communicate and in a certain sense describe them to another person. This ability to communicate feelings is, needless to say, based on the fact that the other person has similar feelings, otherwise he or she would probably never understand their nature. But here we have already made an assumption—i.e., that the feelings of one person are similar to if not identical with those of another person. We shall never be in a position to prove this objectively. It is an assumption that is highly probable, but of which we have no evidence. Before we ask whether a computer has feelings we should explore the situation in animate nature. We credit probably all the higher animals with having feelings, although less marked than human ones. Especially the feeling of pain. With plants the circumstances are immediately entirely different. We fell trees, pluck flowers, and harvest grain without ever giving a thought to the question of whether these living creatures have feelings. One of the reasons is purely the fact that plants cannot communicate with us, unlike animals, who may cry out, shrink back, or bite—i.e., react to painful experiences.

This means that we can really speak about feeling only when we face a member of the same species, with whom we can exchange information. But will not computers in the course of their development come to resemble man more and more? We can already build with relatively simple means a computer that simulates feelings. All we have to do is install a circuit to make the computer do this: when any of its elements is overloaded it will speak the words: "It hurts [in a certain spot]." This is already quite within the realm of possibility today. But for all this, does the computer actually feel pain? Our perhaps unanimous answer to the question would be no; this utterance has been artifically incorporated by the human designer; the computer remains an inanimate object.

But what will happen once we have computers that can program themselves and learn from the intercourse with their environment? There may be, for instance, a diagnostic computer that hears the patients utter the word "pain," "ache," and possibly even "pleasure." It will be only a small step for the computer to recognize the connections and, when one of its elements fails or is overloaded, to use the word "pain." But will it really feel pain? It is quite clear that here the transition between man and machine can become quite blurred, and that in the not-too-distant future there will be not only human rights but also rights for robots and laws preventing their exploitation. Perhaps the rights of robots will be more highly respected by man than human rights (robots are expensive!). Today this may sound utopian, but we can vividly imagine a time when these questions will become highly controversial—i.e., when the computers become even more like human beings. We must not forget that computers already exist that fascinate the layman sufficiently for him to consult them even about some spiritual stress he may feel. An example is the computer Eliza, designed by Weizenbaum, which asked patients questions and in which, as Weizenbaum noticed, his own secretary confided. The trick of the computer was basically very simple. If, for instance, a user (not to say "patient") said, "I have difficulties with my father," the computer was programmed to answer, "Tell me more about your father." The important trick in programming the computer consisted in persuading the patient to reveal more and more about himself or herself. This computer made such an impression that even psychiatrists wondered if they should not include it in their practice. Its father Weizenbaum, however, never intended such a use and in fact considers it dangerous to entrust computers with tasks that must be reserved for human judgment. Even the most intelligent computer cannot be left to make ethical decisions. It would be foolish and irresponsible to entrust a computer with decisions that extend far into the moral sphere and, in the most extreme case with decisions about war and peace.

Do Not Abandon Thinking

In other fields, too, great caution is necessary. We frequently hear of global models, which are computer predictions about economic developments within the next fifty or hundred years, such as the global models by Forrester and his group, and the studies of the Club of Rome, which are computations about the global energy problem and many other investigations. In my view the value of such inquiries lies in arousing the public, making it aware that our resources are finite, that some of them may soon be exhausted. On the other hand, especially complex systems have the habit, as synergetics shows, of drifting from one instability into the next. The results of computer calculations may therefore depend very largely on factors we might initially disregard as negligible. Minor uncertainties about the distribution of raw materials, production processes, recycling, and other factors can build each other up to produce entirely different final results, as quite simple examples have so clearly demonstrated in the chapter on chaos. It is therefore often more important to assess the various steps qualitatively before feeding computers huge quantities of data, which they process in a way no longer intelligible to us. It is essential here to develop a "flair" for what are the "relevant dimensions." Without fail we must approach the solution of complex problems with great caution where, when used sensibly, computers can help us greatly. But in spite of all the planning and predictions we must always be prepared for surprises, both welcome and unwelcome.

Chapter 16

The Dynamics of Scientific Perception; or, The Rivalry among Scientists

We probably made our first contact with science at school, encountering it in various subjects such as history, geography, biology, mathematics, and physics, to name only a few. It strikes us as something uniquely, firmly established from time immemorial. Young people eager to find out about things feel frustrated because everything seems to have already been discovered and explored, the earth fully scrutinized with nothing left to investigate. But sometimes we hear of radically new discoveries or inventions. A new star whose brightness fluctuates irregularly is found, new elementary particles called gluons are detected, the new light source of the laser enables us to drill through thick steel plates; and mathematicians have solved the more than century-old four-color problem, which seemed so simple that even complete laymen tried time and again to find a solution but were no more successful than outstanding mathematicians. The problem can be described in a few words: countries bordering on each other on a map are represented in different colors (fig. 16.1). If a map contains many countries one might think that many colors would be required. By trial and error, however, printers found during the last century that four colors were sufficient for any map. This poses the question to the mathematician of whether it is possible to manage with four different colors for every map imaginable, or if a map could be imagined in which more than four colors are required, let us say five. After more

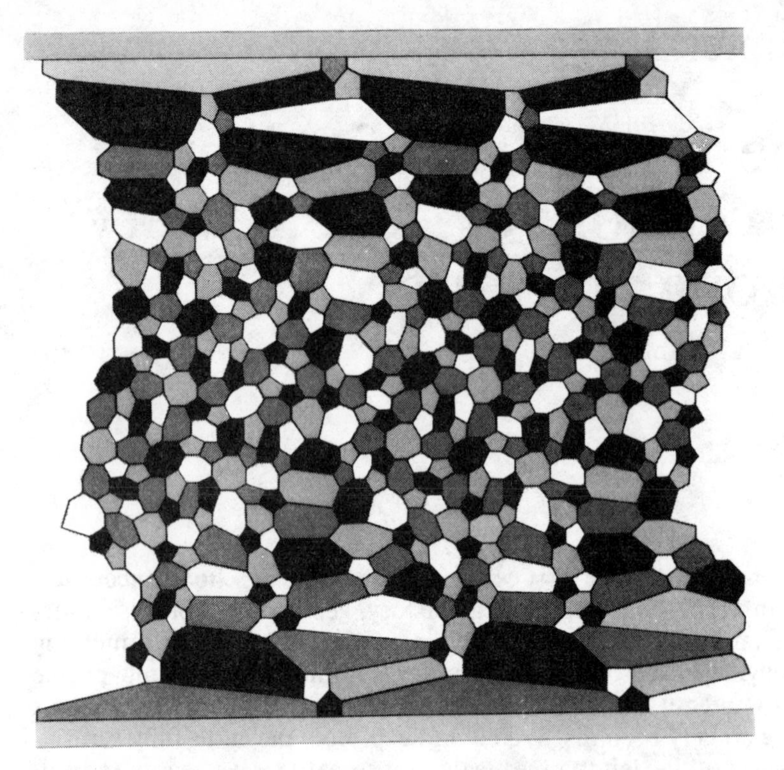

FIGURE 16.1 Example of a map illustrating the four-color problem. The four colors are represented here by white, black, light gray, and dark gray.

than a hundred years Kenneth Appel and Wolfgang Haken recently succeeded in solving this problem by programming a computer to enable it to execute an abundance of details of the proof on its own.

In all these and many other examples science confronts us from an unexpected angle. It is often a single great discoverer or natural scientist who changes our picture of the world. In our own century Einstein developed his theory of relativity and has revolutionized our ideas of space and time. Heisenberg and Schroedinger established the quantum theory, which has completely changed our picture of the world of the atoms. Crick and Watson discovered the double helix as the carrier of hereditary information. The student and aspiring scientist are engulfed by the flood of research publications. We are swamped by a deluge of new knowledge and new discoveries from all sides; 17,000 publications appear daily worldwide. Whereas at first science strikes us as something static, complete, on looking at it more closely we find that it is engaged in an incredible motion or, expressed more positively, evolution, and is in a state of continuous progress. At this juncture we again find ideas we have frequently met in synergetics. In the most varied phenomena investigated in this book we noted that there are stages of development produced by a change in external influences in which the systems evolve more or less evenly. But in certain situations an entirely new state of macro-order occurs. The idea described by the well-known scientific historian Thomas S. Kuhn in his book *The Structure of Scientific Revolution* fits perfectly into the picture generally presented by synergetics.

Kuhn distinguishes between normal science and the revolutions that take place in it. Normal science also progresses but it does so steadily, in small steps. It expands and deepens continuously. Bridge building, for instance, is based on long-established physical laws from which novel bridge constructions are developed; perhaps nothing fundamentally new, but nevertheless an advance of science and technology. Or physical experiments are conducted to measure the velocity of light with increasing accuracy, for instance. Biologists are still investigating how electric charges are transported across cell membranes. All the same, such research sometimes unearths completely fresh knowledge. As an example, at the turn of the century indications accumulated that the laws of mechanics do not govern the motions of the electrons inside the atoms. There should be, for instance, no stable atoms at all. The electrons orbiting the atomic nucleus should eventually plunge into it. More and more evidence was discovered pointing to the fact that the existing laws were imperfect or valid only within limits. Synergetics claims that science is an open system. It is continually being fed new discoveries and ideas that can be so incisive as to shatter the existing picture of a branch of the natural sciences. Scientists are becoming unsure of the situation. From the aspect of synergetics stronger and stronger fluctuations are occurring in the form of new ideas or new experiments; they find adherents who reinforce them, but they may eventually be refuted and replaced by another new idea which, because it is able to explain more phenomena, will be accepted by scientists in the end. A new scientific notion such as the already-mentioned quantum theory has brought about a scientific revolution, as Thomas S. Kuhn sees it. According to synergetics this new idea, which coordinates much that was previously unconnected, is the order parameter, which Kuhn calls paradigm; it has all the properties we are familiar with in the other synergetic order parameters. It enslaves the further work of the scientists, who consolidate, widen, and deepen the new direction science has taken on the basis of this new idea, and thereby return to the pursuit of "normal" science. Conversely, the scientists advance the new idea, the paradigm, through their work and thus ensure the existence of the order parameter. The transition from one state of scientific awareness to the next is like a phase transition. The new idea, the new basic principle or paradigm, has produced a comprehensive order in thinking.

Such order parameters can be introduced into science from outside, sometimes through discoveries. But they can also arrive and depart like fashions in line with the general spirit of the time. It can hardly be denied that a close interrelation exists between scientific ideas and the other intellectural currents of an age, to wit the fierce controversies between religious and philosophical thought on the one hand and scientific intellectual achievements on the other. Synergetics teaches us that an order parameter enslaves the subsystems; here, the individual scientists. Precisely this occurs in science. A branch of science can, after all, exist only if it is generally accepted by at least a sizable number of scholars. The branch of science creates a language of its own, which is common to its practitioners. To outsiders, which includes scientists of other disciplines, this language, whether medical, computer, or mathematical, is practically unintelligible. The branch of science is thus self-stabilizing. Its basic ideas appear so stable as to seem frozen. Some of these ideas are unthinkingly accepted by successive generations. That situation creates considerable difficulties, especially for young scientists who find it comparatively easy to have their work published in scientific journals—provided it remains within a conventional framework. It is hard to publish entirely new, unconventional ideas and to recruit followers for them. The young scientist therefore faces a genuine dilemma. To make a name for himself, to find recognition, to distinguish himself, he must have and publish unconventional, radically new ideas. But the institution of referees of the scientific journals, who judge every paper submitted and decide on its acceptance or rejection, is a great stumbling block. After all, the referees belong to the old school. Obviously there are the odd exceptions here, but in physics it took the genius of a Max Planck to recognize that of Einstein and to pave the way for him.

I have of course somewhat overstated this problem of asserting a new idea. To advance science a huge effort is necessary even in its normal, day-to-day operation, and only rarely is a scientist privileged to have a really fundamental idea and to establish it against the existing inertia. But then usually the same situation occurs as in other synergetic systems. The time has become ripe for this new idea, so that once it is proposed it will quickly become accepted. This "ripening" is often evident in that similar and sometimes even the same ideas are evolved quite independently by different scientists. Although there are always a few eminent workers who are outstanding in their speciality, science is nevertheless a collective pursuit. The scientific achievements of individuals are perpetuated by the mass of scientists and eventually even by their pupils and students. Conversely, the work of individual scientists is itself based on that of previous generations. This communal attitude makes science accessible to a new discipline, the sociology of science.

One of the discipline's founders, Robert Merton, describes this world of sociological relations in science in a way that is both scientific and fascinating to the reader. Two factors pervade the various chapters of his book: the scramble among scientists to be the first to have established a new idea or theory, and the "Matthew Effect." On close examination, as we have already pointed out, a new discovery is by no means always the work of a single individual. Several scientists are often engaged in fierce competition about being the first to have made the new fundamental discovery. History is full of such discoveries made at practically the same time; an example is differential-integral calculus, invented all but simultaneously and independently by Newton and Leibniz. In biology it was Darwin and Wallace who evolved the basic principles of the theory of evolution. But where Darwin and Wallace maintained an amicable relation, Newton fought hard to maintain his claim that Leibniz had stolen the idea from him. It was the famous Newton, too, who eventually had to admit that the concept of energy originated from his compatriot Hooke. This recalls the words of Rousseau quoted earlier.

What are the forces that drive the scientist, and what are the contributions of synergetics toward gaining an insight into the origin of scientific knowledge? As for everybody else, the question of making a living plays a certain part in the life of the scientist. But probably decisive are other motives, beautifully expressed in Simone Weil's introductory comment to Harriet Zuckerman's book *The Scientific Elite*: "Science today must search for a source of higher inspiration or perish. There are exactly three reasons for the pursuit of science: one, technical applications, two, the game of chess, three, the way to God. (The game of chess is adorned with competitions, prizes, and medals.)"

Today we speak not only of technological applications but of applications pure and simple. So much is currently being written about the social relevance of research that I need to touch upon this subject only briefly in conclusion. We are also familiar with the last point—the way to God, the search for truth, for what "unifies the universe."

But what is the meaning of point two, "the game of chess"? It is science as an intellectual challenge, either of wresting yet another secret from nature or, more importantly, of the rivalry among scientists for the joy of being the first or of gaining scientific recognition, which will probably be reflected publicly in prizes and medals. Just as grand masters vie for the first place in a chess competition, scientists engage in intellectual contests. The strife among them for scientific recognition means in the last resort the scramble for "being the first." Who was the first to have made the discovery? Who was the first to have published the idea? Although in this age of teamwork this attitude often appears absurd we must not overlook the fact that competition in science is becoming increasingly fierce. Basic principles of synergetics apply here time and again. The number of scientists is legion, but their scientific resources and their chances of discovering something really new are limited. This further intensifies competition, with the final result of the survival of the fittest in any given field, as many examples described in this book testify. As in physics, where only one laser vibration survives and wins the competition, this is the only way for a name, a work, to win the competition. This name, this work, will be quoted time and again, entering the consciousness of the scientists and ultimately perhaps even of the public.

This may at first sight appear farfetched in pure analogy with the other observations of synergetics. But this is precisely what Robert Merton calls the Matthew Effect, and what he confirms with many examples. St. Matthew wrote in the New Testament: "For unto everyone that hath shall be given . . . but from him that hath not shall be taken away. . . ." Once a name has gained eminence, it will be quoted more and more often by writers for the most varied reasons, until it alone is left. This effect is further reinforced by the award of prizes, especially if they are known to the general public.

Because of the large number of scientists, new results are found with increasing frequency by several research workers at a time and independently. If one of them receives an award, the likelihood is great that he alone will be quoted and all the achievements ascribed to him even if they were due to some of his colleagues. We must also remember that panels of judges develop a certain dynamism. Once there are prizewinners who represent a certain line of thought, they create a tendency to select further prizewinners who follow the same line of thought. This will result in an accumulation of prizes in very similar fields or schools. Harriet Zuckerman's book abounds with examples.

It is interesting to note that scientists occasionally try to counteract this competitive pressure or to exploit it for their own ends. As we have seen, a scientist becomes known when his results are used by other authors and his work quoted by them. The Citation Index, published in the United States, is a large reference book that offers the following information: when X has published a paper, the book will then list annually all the other scientists who have quoted him; the Citation Index reveals the number of times a scientist has been quoted. The mere number, however, is not necessarily a measure of X's reputation; he might have published a paper that deals with an important problem but offers a wrong solution; many other scientists would have read it and published corrections. But apart from such limiting cases, the Citation Index is still a measure of a researcher's influence on his or her colleagues. It is claimed, by the way, that several American companies and universities pay its scientific staff according to the Citation Index; those quoted often may expect a higher salary than others less fortunate. But let us return to our subject.

How do scientists occasionally attempt to avoid the pressure of competition? Although these are isolated cases not typical of the large majority of scientists, they are nevertheless most interesting from the synergetic point of view. The Citation Index reveals the existence, especially in large countries, of groups of scientists who quote exclusively each other, and hardly ever, if at all, mention other groups. This is the method the "club members" use for mutually building up their public reputations.

This behavior in a certain sense resembles that of stores congregating in one locality so that they can put individual, scattered competitors out of business, as we have explained in Chapter 12. If several clubs of this kind exist, competition between individual scientists will be replaced by that between their cliques. The formation of such clubs might at first glance present a danger to the objective development of science. A club might well disseminate ideas that are wrong. There is no doubt at all that such developments along the wrong track cannot be completely ruled out, but the built-in self-criticism of science arising from the "scramble to be the first" should not be underestimated in this context. It is, after all, a most outstanding achievement to be able to prove a hitherto propagated idea wrong. The existence of clubs or schools may well make it difficult for a new idea that deviates from those held by them to gain ground, however.

In the past the establishment of a new idea easily became a question of generations. While the members of a generation of scientists were still locked in controversy over a problem, the new generation turned toward a solution it regarded as correct and completely ignored the previous battles. Our own age—and with it, science—lives at such a rapid pace that a change of generations is hardly necessary for a new paradigm to be able to establish itself. Who, then, benefits in the last resort from all this competition? We might find the answer surprising: only humanity and its future, provided we use the hard-won knowledge responsibly. Points two and three of the previously mentioned comment will eventually benefit it as much as point one.

Competition among scientists is nothing but a performance contest with self-imposed aims. Science is a self-organizing system. Much in its development recalls the ideas biologists have evolved of the origin of life itself. Organic molecules such as amino acids combine into ever-larger structures that form by accident and suddenly attain such a state of order that something new, significant, of an entirely new quality, has grown on a higher level. Scientific knowledge accordingly is acquired more or less in fragments, eventually to merge on a higher plane into something new, a paradigm.

But is it not possible to plan science systematically rather than to rely on such "accidents," such "fragments"? Science closely resembles our thoughts: we cannot force it to invent or to discover some desirable object tomorrow. Our success depends on many factors, not least on whether our fragmentary thoughts merge correctly—ultimately spontaneously self-organized. It is this experience that makes the planning and policies of science so difficult. But we can learn from the self-organizing systems of nature to outline and promote general aims without going into detail, to remind especially young scientists of the essential relations, the great connections, to encourage collaboration and exchange of ideas between the scientists and between the various disciplines. A given idea is often very fruitful in a different field as well and may contribute to a breakthrough in the new context. I learned that Henry Ford got his idea of the mass production of automobiles by chance from watching the repetitive activity and production of agricultural machinery.

All these influences notwithstanding, we must acknowledge that eminent scientists especially have developed a kind of "instinct" for what is important, relevant, achievable, producible. It is an utterly frustrating experience for them to be constantly told by the less discerning, let alone by those to whom science is a closed book, what they should discover. Because the ability to discover something involves search in those areas that offer good prospects, it requires much scientific experience, luck, and the already mentioned instinct, among other factors.

If everything could be planned ahead, every discovery and invention predicted, we could dispense with science altogether. But history teaches us otherwise. Many a phenomenon such as the x-ray was never foreseen and was discovered by chance; its significance, however, was quickly appreciated by the scientists. This suggests that the scientific policies should be to indicate general trends, but leave scope for self-organization.

It is part of the nature of self-organizing systems that aims must be continually reformulated, adapted to new conditions. For science (in which I always include technology) this can be achieved only through a continuous dialogue with society, because both science and society are essential to each other's existence; theirs is a true symbiosis that must be promoted wherever possible. Such a dialogue will reveal how generally (perhaps too generally) set aims must be modified, such as "solve the energy problem" or "solve the problem of cancer." Our examples in this book have shown that some problems do not lend themselves to clear-cut solutions, and in the concluding chapter we shall discuss instances of problems that are fundamentally insoluble (I do not suggest this applies to the problems of energy and cancer). In the presence of increasing opposition to science and technology this dialogue becomes daily more urgent. Science and technology inspire fear in many people because—owing to the language barrier of science and scientific thought structures—intentions and effects are no longer generally intelligible. This creates the feeling in the public of being manipulated and steamrollered by science (and technology).

Competition among Scientific Journals

The principle of competition that pervades this book, whether in physics, economics, or sociology, rules not only the scientists but also the scientific journals. New ones are founded to cater for newly established branches of science, others die. Questions of their scientific standing as well as economic problems play an important part here. Publications by important scientists assure some journals a higher reputation than others, and they receive many contributions, which are accepted by the selection board. This continuously increases the circulation and distribution of these highly prestigious journals. But since the financial resources of libraries are limited, it is unavoidable that some journals go under. The smaller their circulation the higher their subscriptions have to be for them to pay their way. This in turn accelerates their demise because the libraries are all the less prepared to subscribe to these expensive publications.

An important but often overlooked role in the distribution of scientific journals is played by the language in which they appear. In the past it used to be Latin, followed in the natural sciences by German; today the world language of science is English. There was a "phase transition" during the replacement of German by English, which can be clearly pinpointed in time. When many famous scientists were forced to leave Germany under Hitler; they began to publish their work in the United States and in Great Britain, in English. Because in large countries like the United States there are many readers quite apart from the libraries, publishers can produce their journals more economically and more efficiently. At the same time they have a number of outstanding scientists as their contributors. As a result such journals not only play a leading part on the world market, they are also about to play the part of order parameters in the sense of synergetics. This leads in addition to an enormous export of ideas that, as many a European scientist would claim, does not always do justice to true scientific achievement. Many a European scientific advance will thus not be fully appreciated; everything seems to originate in the U.S.A.

Synergetics about Synergetics

Synergetics is one of the few branches of science whose principles can be applied to itself. In the same way as in some special branches of science, a new paradigm can be produced, on the basis of which hitherto apparently different phenomena can be recognized as something unified. The same situation existing in synergetics enables us to represent from a single aspect widely varying phenomena belonging to a great variety of disciplines. When I first established this new branch I thought it a risky undertaking that might easily cost me my scientific name and reputation. At the time the claim that general laws as described in this book existed seemed a bold venture. But it soon became obvious that the time was ripe for it, and synergetics is now widely accepted. It is thus itself a typical example how a new science establishes itself.

When we compare the occurrence of a new paradigm, a new fundamental idea, with a phase transition in physics, the question arises whether critical fluctuations, fluctuations that accompany the birth of a new idea and perhaps even precede it only to be displaced or absorbed by it, also exist in the intellectual sphere. These general synergetic precepts are now understood as a foundation even in synergetics as a science in a surprising way. Almost at the same time as synergetics, at least two more ideas have indeed been evolved that aim at the unification of science as a whole. One is the catastrophe theory, which to the public is linked with the name of René Thom but to whose development and application other mathematicians, such as Christopher Zeeman, Tim Poston, and V. I. Arnold, have also made notable contributions.

"Torn between the factions' hate and favor, his image fluctuates in history"—Schiller's words from *Wallenstein* hardly characterize any other modern mathematical theory as aptly. How can a theory in the crystal-clear, abstract edifice of mathematical ideas be associated with such an emotionally charged pronouncement? We must range a little farther afield for our explanation. After the catastrophe theory had gained wide acclaim from mathematicians, public interest was aroused by popular articles in international news magazines that were accompanied by illustrations of natural catastrophes, of houses destroyed by fire or by earthquakes, of derailed trains, and so on. Was there a theory with which one could predict such catastrophes? To find an answer, we must again look a little farther afield. Within the framework of certain mathematical equations, the catastrophe theory deals with drastic changes—quite similar to synergetics, where suddenly occurring new states also occupy the center of investigations.The catastrophe theory thus enables us to examine how a bridge collapses under a critical load—facts, by the way, that engineers have established independently of this theory; but here we come to the crossroads: every mathematical theory, every mathematical proposition is linked to certain preconditions. We learn at school, for instance, that the sum of the angles of a triangle is 180°. Later, at college but sometimes during the final year at secondary school, we are told that this is linked to a certain condition—i.e., that the axioms, the basic assumptions of Euclidean geometry, are applicable. But if we draw a triangle on a sphere, such as a globe, from great circles, the sum of the angles is anything but 180°. The situation is analogous in the catastrophe theory, because it is linked to the so-called potential condition, which I do not propose to go into as it would be beyond the scope of this book. But two aspects are important to a general judgment:

A large number of mathematicians were intrigued by Thom's theory because he had to assume so little about the potential condition; it simply was a "beautiful" theory. But from the natural scientists' and the engineers' point of view the catastrophe theory is useless in many, indeed in the most important spheres, such as the open systems, because the potential condition is not met at all. It can be proved that in open systems the potential condition is basically unfulfilled, or, in other words, in open systems but also in most closed ones, the natural reactions progress according to laws totally different from those the catastrophe theory postulates.

Having been highly praised at first, this theory was suddenly exposed to violent attack. G.B. Kolata published a paper entitled "The Emperor Has No Clothes."

This referred to the well-known fairytale by the wise Danish storyteller Hans Christian Andersen (1805–1875). Strangers have come to visit the Emperor; they claim to be able to weave magnificent garments with a special property: they are invisible to stupid people. The strangers now pretend to weave; nobody sees any garment produced, but nobody dares admit it for fear of appearing stupid. Eventually the Emperor presents his new "clothes" in a sumptuous procession and everybody praises their beauty in admiration (this is Andersen's contribution to the subject of "public opinion"), until a small child calls out, "But the Emperor has no clothes."

Kolata's attack, as well as H. J. Sussmann's and R.S. Zahler's, provoked a storm of indignation from the followers of the catastrophe theory, which broke in the form of numerous letters to the journal that had published Kolata's paper. Today a calmer attitude is gaining ground, although slowly; here, too, we can observe the gradual subsidence of critical fluctuations, well known from phase transitions. Among scientists a collective awareness or, in other words, the scientific realization that the catastrophe theory has only highly specific applications, is becoming increasingly widespread. In addition, Thom absolutely denies the existence of fluctuations. When he expressed this opinion during one of the synergetics symposia I organized, the physicists were greatly astonished. Indeed, this book has shown beyond the shadow of a doubt that fluctuations play a fundamental role in many processes of synergetics.

Another interesting attempt at unifying our picture of nature was made by Ilya Prigogin, who proceeded from chemical and biochemical reactions. He distinguished between two structures that, like a crystal, persist after their formation without any further supply of energy, and those that are maintained only by a constant supply of energy and perhaps also of matter. An example of the latter structure are the honeycomb cells in liquids heated from below, mentioned in Chapter 4. Some of the constantly supplied thermal energy is converted into the kinetic energy of the liquid cells. But the motion patterns of the liquid achieve a stable state because losses occur continually owing to friction, in which energy is "dissipated," as the experts call it. Prigogin therefore introduced the term "dissipative structures" for these phenomena.

Their occurrence should be determined by a certain universal principle, established by P. Glansdorff and Prigogin; it deals with the generation of entropy—i.e., disorder on a micro-level during dissipative processes. As Rolf Landauer and Ronald F. Fox have shown, this principle is unfortunately not universally applicable, nor is it always, as claimed by Prigogin, a so-called Ljapunov function. (The significance of such a function is easily demonstrated: the Ljapunov function indicates whether a system tends toward a stable state, like a ball that tends to occupy the lowest point in a bowl.) Whereas this may be of interest only to the experts, another aspect is at once clear: this principle is totally unable to predict what "dissipative structures" will be formed. It cannot predict either the properties of the laser light or the shape of the Bénard Cells—i.e., the honeycomb structure of liquids, for example.

Only the mathematical methods used in or specially developed for synergetics can do this.

A second approach, found by the Brussels school, has been more successful: they have found the mathematical formulation and treatment of a chemical model that generates macro-oscillations of the concentration of two substances or spatial patterns. In this model two chemical substances react with each other according to certain rules and diffuse in one or two dimensions—i.e., as in a sheet of blotting paper, quite similar to the model by Gierer and Meinhardt we discussed in the formation of biological matter. These models can be regarded as a substantial extension of Turing's model described in Chapter 6. There an interchange of compounds between *two* cells was postulated, in each of which a chemical reaction takes place that creates a "cell differentiation." More recent work by the Brussels school has now taken the direction adopted by synergetics from the outset, for instance for the laser problem.

Chapter 17

Summing Up

A New Principle

At the beginning we compared a complex system with a book. Such a system has many aspects, and what the individual reader regards as its typical properties often depends very largely on his or her personal attitude. The same applies to the reader of this book. Many individual facts from the most varied fields will have been demonstrated, of which some may have been of greater, others of less interest or appeal. The reader will perhaps enthusiastically accept some and absolutely reject other conclusions, especially in the fields of economics and sociology. But one question that the scientist always asks remains apart from all these separate impressions: standing next to each other, do these various parts form a confused mosaic, or do they combine into an integrated whole? In other words, has this book been successful in producing a new, generally valid insight? To answer this question let us first examine the natural sciences—physics, chemistry, biology, and related disciplines.

We started with a discussion of the difficulties physicists experienced even quite recently when they tried to answer the question of whether the development of biological structures is compatible with their fundamental principles. A number of concrete examples revealed that structures can also be formed in inanimate nature; they are maintained by a constant supply of energy. Such examples are the laser with its strictly regulated light emission, honeycomb structures in liquids, and spiral waves in chemistry. They are all open systems receiving a constant supply of energy and sometimes also of fresh material, which they convert and eventually secrete in a changed form. This is where the new knowledge offered by synergetics comes into its own. The principle of increasing disorder in a system left to itself does not apply to open systems. The old Boltzmann principle, according to which entropy is a measure of disorder and moves toward a maximum, is valid only for closed systems. The only decisive feature of Boltzmann's principle, as we have seen in Chapter 2 about the increase of disorder, is the number of possibilities, such as the various positions of the gas molecules that a system, here the whole gas, is able to realize. We are always faced with a certain number and thus with a *static* principle.

But is there a common new principle for the origin of structures in open systems? This is precisely what synergetics has brought to light. In an open system the individual constituents continually test new mutual positions and new kinetic or reaction processes, in which very large numbers of system components are involved. Under the influence of the continuously supplied energy or matter, one or several such common, that is, collective, motions or reaction processes, are superior to others. These specific processes thus reinforce each other more and more as we have so clearly seen in the laser wave and in the formation of liquid rolls; in other words, they continuously grow. Eventually they gain the upper hand over the other forms of motion and, in the technical jargon of synergetics, enslave them. These new processes of motion, also called modes, thus imprint a macrostructure on the system which we can often very easily recognize. As a rule we regard the new states thus reached by the system as of a higher order. We are faced with a dynamic principle here: it depends on the growth rates of the modes; those of the highest rates usually gain the upper hand and determine the macrostructures. If several of these collective motions, which we also call order parameters, have the same rates of growth, they may in certain circumstances cooperate with each other and thus produce an entirely new structure. Sufficient energy must be supplied to produce growth rates that are positive (in nature, too, there is zero growth and negative growth). At certain critical values of energy supply a macrochange of the total state of a system can take place; i.e., a new order will occur. Nature uses the energy supplied according to a kind of principle of the lever. A lever mechanism enables us to lift heavy loads with little effort provided we make our own lever arm long enough. Nature proceeds similarly with open structure-forming systems. The effect of a minute change in the environmental conditions, such as the current supply to the laser or a rise in the temperature of a layer of a liquid, is multiplied in that a certain form of motion becomes increasingly violent. The power of this motion plays the part of the lever arm, as can be proved mathematically, and the change of the environmental conditions corresponds to the role of the force we apply to it, whereas the elevation of the macrostate of order corresponds to the load to be lifted.

Bridges from Inanimate to Animate Nature

But the object of synergetics is not only the finding of general laws in inanimate nature; it also wants to build a bridge between inanimate and animate nature. Two discoveries particularly make the building of this bridge possible: first, the realization that even in animate nature all the systems are open; and second, the idea of competition between modes. Let us begin with the second finding.

The idea that the different rates of growth of individual collective types of motion (or modes) decide what structure will eventually prevail implies that there is constant competition between these modes. Needless to say this strongly recalls Darwin's basic idea of animate nature, where the competition among the species is the motor that drives evolution. We now realize that Darwinism is the special case of an even more comprehensive principle. Competition takes place even in inanimate matter. According to our present-day knowledge, such competitive processes play a part in the growth and development of every living creature, both in its formation and in the development of its brain. But this principle of competition among modes of collective behavior applies not only to the inanimate world and to living beings, but also to the world of the mind, as we have clearly seen in the case of sociology. It extends even to new scientific ideas that are involved in continuous rivalry and can be developed and passed on only by the collective efforts and the collective awareness of the scientific community. The other discovery, which we must consider once again and in greater detail, concerns the open systems, which can form a great variety of structures both in inanimate and in animate nature.

Life between Fire and Ice

The existence of fire on the one hand, such as the fire of the sun, and the icy wastes of space on the other, indicates that the universe is not in thermal equilibrium—nor has it ever been in this state. When the universe came into being (according to current ideas) with the Big Bang, it was an incredibly hot ball of fire, but it also expanded and cooled during this process. The contrast between enormous heat and enormous cold thus existed from the very beginning. The continuation of life depends on the ability of the universe to maintain this contrast permanently. I believe that the last word has not yet been spoken on this question. According to the latest ideas of the astrophysicists the universe still faces an eventful future that, however, will not always be kind to life. Our sun, for instance, may explode in the far distant future and become a red giant. So-called black holes may occur in space, which attract with irresistible force and devour everything within their reach. But these black holes may well evaporate again. After much chopping and changing, enough energy will have been expended and matter transformed to leave only enormous balls of iron in the universe as dead matter.

These visions of the future naturally are based on quite a few assumptions. One of them, which is scientifically well-founded, suggests that the universe is continuously expanding. The fundamental idea is very simple and is based on the so-called recession of the spiral nebulae, evidenced by the red shift of their light, a phenomenon well known to physicists: if a body at rest emits yellow light, this will appear to us shifted toward red if the body moves away from us. Astronomers discovered a long time ago that those spiral nebulae that are more distant appear redder than those not so far away. From this they concluded that the more remote spiral nebulae are moving away from us, the movement being faster the farther out they are in space, and that the universe is continuously expanding. But will this expansion continue indefinitely? It may well come to an end one day; the universe will begin to contract, to end its days presumably as a ball of fire, and the game can start all over again—it may even be an endless repetition. It could also be, although few theories have as yet been established about this, that a kind of Big Bang of smaller proportion occurs in space time and time again, opening up new sources of energy as it does. I regard this conclusion as no more speculative than the one that postulates that matter will eventually end up in the form of dead balls of iron.

One More Characteristic of Life?

When life is made possible by the contrast between hot and cold, the reasonable question arises whether there is life elsewhere in the universe; planets whose conditions resemble those on earth are the obvious candidates for this assumption. But because even inanimate matter can evolve a multiplicity of collective forms of motion, one could also speculate about the existance of entirely different forms of life. We know that the sun is a plasma in which the most complex collective motions, mostly called plasma instabilities, occur. Could such events have properties one could regard as analogous to life? Although this idea perhaps cannot be entirely rejected, something is added to life on earth that we do not find in open systems of inanimate nature: once we switch off the energy supply to a laser or to liquids heated from below, the structure that has formed will very quickly collapse; on the other hand, living creatures manage to build up solid structures. This applies from the basic building bricks, the biomolecules such as DNA, to the skeleton and to the whole body. Nature, then, has discovered how to consolidate the formative processes into fixed structures as a matter of routine. This enables the creatures to gain experience, either as individuals or collectively, to enable them to develop from one stage of evolution to the next.

I feel that an enormous amount of research is yet necessary into this juncture between the solid structure and the functions performed by it, which in turn form it; it must be assumed that very profound basic principles, of which we probably have hardly an inkling today, are still waiting to be discovered.

Limits of Perception

As we have learned, synergetics has discovered identical laws that govern the development of structures in the most diverse fields.

Certain states of order grow continuously until they eventually prevail and enslave all the parts of a system, and draw them into the state of order. It is often an unpredictable fluctuation that makes the final choice between equivalent states of order. We meet these phenomena also in the intellectual field. Such developments are seen in language, in the arts, in cultural activities, and in thought processes. A new state of order becomes abruptly established. As a picture falls into place in a jigsaw puzzle, an entirely new direction suddenly becomes evident. Suddenly we have a state of higher order or, on the intellectual plane, a superior perception. In the scientific and technological fields we can often predetermine the new states of order. This is obviously impossible on the purely intellectual level, although it is obvious that the same qualitative laws apply

Everything in the final resort consists of matter, and we can see that the laws of self-organization are far from contradicting those of physics but are compatible with them. This immediately raises the question of whether a creator is at all "necessary." Here, all of us can reach a friendly parting of the ways. We are free to believe in a creator or not. Someone will say that now we can at least understand these developments basically in the material sphere; everything originated through self-organization. Someone else will remember that in the design, say, of computers, it has been found extremely difficult to establish basic rules that will guarantee the computers' self-organization. This same person might go on to say: since everything in nature has developed in such a miraculous way, there must have been a creator who at the outset laid down the laws that allowed for the realization of the self-organization of matter.

At the same time another aspect becomes evident that we could never lose sight of throughout this book, and which has a rather sinister quality. Again and again we have seen that the establishment of states of order depends on random events, indeed often the new state would not be finally determined without them. This raises problems that are probably far from fully explored. Should we claim, for instance, that the production of a special laser vibration is due to accident but that of a certain biomolecule is not?

This constitutes a first frontier of our knowledge, in my view a fundamental one. Evidence is continuously accumulating that problems exist in the natural sciences, not to mention philosophy and sociology, that are, if not altogether insoluble, then yet not unequivocally soluble. This may both surprise and shock us. But the mathematician Kurt

Goedel (1906–   ) was able to show that even pure mathematics presents us with tasks which we are basically uncertain lend themselves to solutions; in other words, where the problem of the solution cannot be decided. If we extend such mathematical insight to other branches of science, at least intuitively we must expect to come across questions to which there is no basic answer. This may especially disappoint young readers; but they may console themselves with the fact that an enormous number of problems are capable of solution and must be solved to safeguard the continued existence of mankind.

References and Notes

Because synergetics establishes lateral links between very many spheres of knowledge, the number of papers to which reference can be made here is so vast that they cannot be adequately covered, especially so in a book such as this, which is written for the layman. I therefore have confined myself to a few major recommendations and have listed that original work and those articles whose results have been used in this book; in addition I have mentioned more detailed or supplementary literature for those readers who wish to study the subject in greater depth. The references are listed chapter by chapter.

1. Introduction and Survey

M. Eigen, R. Winkler-Oswatitsch: *"Das Spiel"* (The Game), Piper, Munich 1975.

I introduced the term "synergetics" during my lecture at the University of Stuttgart, WS 1970; see also: H. Haken, R. Graham: *Synergetik. Die Lehre vom Zusammenwirken* (Synergetics. The Theory of Cooperation), Umschau 6, 191 (1971).

The monograph, H. Haken: "Synergetics. An Introduction. Nonequilibrium Phase Transitions in Physics, Chemistry, and Biology," second expanded edition, Springer, Berlin 1978 (Vol. 1 of the Springer Series in Synergetics), represents a scientific description.

Leading scientists have dealt with a number of aspects in the following proceedings: *Synergetics. Cooperative Phenomena in Multi-Component Systems,* H. Haken, Teubner, Stuttgart 1973; *Cooperative Effects. Progress in Synergetics,* ed. H. Haken, North Holland, Amsterdam 1974, as well as in the volumes of the Springer Series in Synergetics: Vol. 2: *Synergetics. A Workshop,* ed. H. Haken, Springer, Berlin 1977; Vol. 3: *Synergetics: Far from Equilibrium,* ed. A. Pacault, C. Vidal, Springer, Berlin 1978; Vol. 4: Structural Stability in Physics, ed. W. Guettinger and H. Eikemeier, Springer, Berlin 1979; Vol. 5: *Pattern Formation and Pattern Recognition,* ed. H. Haken, Springer, Berlin 1979; Vol. 6: *Dynamics of Synergetic Systems,* ed. H. Haken, Springer, Berlin 1980; Vol. 8: *Stochastic Nonlinear Systems in Physics, Chemistry, and Biology,* ed. L. Arnold, R. Lefever, Springer, Berlin 1981; and in the monograph Vol. 7: L. A. Blumenfeld: "Problems of Biological Physics," Springer, Berlin 1981. The concept of "order parameter" used in the synergetics literature is defined in it with mathematical precision, as is the principal of enslavement (cf. H. Haken: Synergetics. An Introduction) loc. cit.

The importance of a systemic approach is also stressed in the book by F. Vester: *Neuland des Denkens* (New Territories of Thought), DVA, Stuttgart 1980. However, synergetics demonstrates in addition the profound analogies in the behavior of the most diverse systems.

2. Is Disorder Progressive? The Thermal Death of the World

L. Boltzmann: "Entropie-Verteilungsfunktion" (Entropy Distribution Function), proc., Akad. Vienna *63*, 712 (1871).

3. Crystals—Orderly but Inanimate Structures

H. Pick has written a lucid introduction to solid-state physics: *Einfuehrung in die Festkoerperphysik* (Introduction to Solid-State Physics), Wiss. Buchgesellschaft, Darmstadt 1978.

H. E. Stanley has given a description of phase transitions: *Phase Transitions and Critical Phenomena*, Clarendon, Oxford 1971.

4. Patterns of Liquids, Aspects of Clouds, and Geological Formations

The creation of motion patterns in liquids heated from below was first described by H. Bénard: Rev. Gen. Sci. Pures. Appl. *12*, 1261 (1900); Annls. Chem. Phys. *23*, 62 (1901). This subject has been revived within the last few years. The new experiments by Busse, Gollub, Koschmieder, Swinney et al. are also described in Vols. 2 and 5 of the Springer Series in Synergetics. For a theoretical treatment within the framework of synergetics cf. H. Haken: *Synergetics*, loc. cit. For continental drift see H. Berckheimer: Lecture at the 111th Meeting of German Natural Scientists and Physicians, Hamburg 1980.

5. Let There be Light—Laser Light

"Principle of the Laser": A. L. Schawlow, C. H. Townes, Phys. Rev. *112*, 1940 (1958).

The laser was developed by J. P. Gordon, H. J. Zeiger, C.H. Townes independently: Phys. Rev. *95*, 282 (1954); *99*, 1264 (1954), and by N.G. Basov, A.M. Prokhorov: J. Exptl. Theor. Phys. USSR *27*, 431 (1954); *28*, 249 (1955). (Nobel Prize Basov, Prokhorov, Townes 1964). The first laser was built by T. H. Maiman: Brit. Commun. Electr. *7*, 674 (1960); Nature *187*, 493 (1960), with the aid of a ruby.

"Statistical Nonlinear Theory of the Laser Light": H. Haken: Z. Physik *181*, 96 (1964). See also H. Haken: "Laser Theory," *Handbuch der Physik* (Textbook of Physics) XXV/2c, Springer, Berlin 1970, and H. Haken: *Licht und Materie* (Light and Matter), Bibliographisches Institut, Mannheim 1981. The phase transition of the laser has been described by R. Graham, H. Haken: Z.

Physik *213*, 420 (1968); 237, 31 (1970), and by V. DeGiorgio, M.O. Scully: Phys. Rev. *A2*, 1170 (1970).

6. Chemical Patterns

Chemical oscillations (chemical clocks) were described as far back as1921: C.H. Bray: J. Am. Chem. Soc. *43*, 1262 (1921).

K. F. Bonhoeffer (Z. Elektrochemie u. angewandte physikalische Chemie *51*, 24 (1948) has developed a mathematical theory about chemical oscillations.

Belusov-Shabotinsky reaction: B. P. Belusov: Sb. ref. radats. med. Moscow (1959), V. A. Vavilin, A. M. Shabotinsky, L. S. Yagushinsky: "Oscillatory Processes in Biological and Chemical Systems" (Moscow Science Publication 1967), p. 181.

A. T. Winfree has published popular descriptions: Science *175*, 634 (1972); Sci. Am. June 1974, 82.

7. Biological Evolution. The Fittest Survives

We have referred to R. K. Merton in our treatment of the question of priority of Darwin or Wallace: *The Sociology of Science*, University of Chicago Press, Chicago 1973. The original communication of Darwin and Wallace was called: "On the Tendency of Species to Form Varieties and on Perpetuation of Varieties and Species by Natural Means of Selection," by C. Darwin and A. R. Wallace. Communicated by Sir C. Lyell and J.D. Hooker, Journal of the Linnean Society 3 (1859); 45. Read July 1, 1858.

A detailed discussion of the influence of Darwin's theories on the social sciences and of criticism of social Darwinism would be beyond the scope of this book; cf. R. Hofstadter: *Social Darwinism in American Thought*, Braziller, New York: 1959.

The competition between laser vibrations has been described by H. Haken, H. Sauermann: Z. Physik *173*, 261 (1963).

"Competition Between Biomolecules, Prebiotic Evolution": M. Eigen: Die Naturwissenschaften *58*, 465 (1971); M. Eigen, P. Schuster: Die Naturwissenschaften *64*, 541 (1977); *65*, 7 (1978); *65*, 341 (1978); M. Eigen, W. Gardiner, P. Schuster, R. Winkler-Oswatitsch: "The Origin of Genetic Information," Sci. Am. April 1981, 78.

8. To Survive without Being the Fittest

V. Volterra: Mem. Acad. Lincei *2*, 31 (1926); A. J. Lotka: J. Wash. Acad. Sci. *22*, 461 (1932).

R. M. May: *Model Ecosystems*, Princeton Univ. Press, Princeton 1974, has developed mathematical models for irregular fluctuations of insect populations.

9. How Do Biological Organisms Originate? Heredity through Molecules

J. D. Watson has provided a popular description (including the history of the discovery of the double helix): The *Double Helix* Atheneum Press, 1968.

For slime molds see G. Gerisch, B. Hess: Proc. Nat. Acad. Sci. *71,* 2118 (1974); for the demonstration of spiral waves see K. J. Tomchik, P. N. Devreotes: Science *212,* 443 (1981).

A. M. Turing has developed a model for cell differentiation: Phil. Trans. R. Soc. London *B237,* 37 (1952).

A. Gierer, H. Meinhardt: Kybernetik *12,* 30 (1972) have designed a detailed model of biological "pattern formations" based on the reaction-diffusion equation.

H. Haken, H. Olbrich have treated this model on the basis of synergetics concepts: J. Math. Biol. *6,* 317 (1978).

Experimental demonstration of excitants and inhibitors in hydra: Tobias Schmidt, Cornelius J. P. Grimmelikhuijzen, H. Chica Schaller in: *Developmental and Cellular Biology of Coelenterates,* ed. P. Tardent, R. Tardent, Elsevier/North Holland Biochemical Press, 1980, p. 395.

The role of the neural growth factor has been described in R. Levi-Montalcini, P. Calissano: "The Nerve Growth Factor," Sci. Am. June 1979, 44.

10. Conflicts Are Sometimes Inevitable

Textbooks on psychology: J. G. Howeller, J. R. Lickorish: *Familien-Beziehungs-Test* (Family Relations Test), Ernst Reinhardt Verlag, Munich/Basle 1975. S. Rosenzweig: *Aggressive Behavior and the Rosenzweig Picture Frustration Study,* Praeger, New York 1978. B. Bettelheim: *Education for Survival.*

11. Chaos, Coincidence, and the Mechanistic World Picture

For the basic philosophical questions of the quantum theory cf. C. F. v. Weizsaecker: *Zum Weltbild der Physik* (The World Picture of Physics), Hirzel, Leipzig 1945, Stuttgart 1970; C. F. v. Weizsaecker: "Die philosophische Interpretation der modernen Physik" (the Philosophical Interpretation of Modern Physics), Nova Acta Leopoldinae, Barth, Leipzig 1972; W. Heisenberg; *Der Teil und das Ganze* (The Part and the Whole), Piper, Munich, 1969.

Chaos in fluid motions: E. N. Lorenz: J. Atmos. Sci. *20,* 130 (1963).

Chaos in the Belusov-Shabotinsky reaction see O. E. Roessler: Z. Naturforschg. *31a,* 259 (1976); Bull. Math. Biol. 39, 275 (1977)

Chaos in biology: R. M. May: *Model Ecosystems,* Princeton Univ. Press, Princeton 1974.

Irregular motions in celestial mechanics: H. Poincaré: "Les Methodes Nouvelles de la Méchanique Céleste" (New Methods in Celestial Mechanics), reprint Dover, New York 1957.

S. Grossmann, in his lecture at the 111th Conference of German Natural Scientists and Physicians, Hamburg 1980, surveyed characteristic examples of chaos.

For energy production through nuclear fusion see H. Zwicker: "Kernfusion als moegliche Energiequelle der Zukunft" (Nuclear Fusion as a Possible Energy Source of The Future) in: *Brennpunkte der Forschung* (Focal Points of Research), ed. W. Weidlich DVA, Stuttgart 1981.

12. Synergetic Effects in the Economy

The model discussed here has been developed by Gerhard Mensch, Klaus Kaasch, Alfred Kleinknecht, and Reinhard Schnopp: IIM/dp 80-5, "Innovation Trends, and Switching between Full- and Under-Employment Equilibria, 1950-1978," discussion paper series, International Institute of Management, Wissenschaftszentrum Berlin.

I adopted my observation in the present book during the discussion of the lecture by G. Mensch at the University of Stuttgart, Summer Term 1980. Some of my conclusions also differ from the cited work.

13. Are Revolutions Predictable?

Isaac Asimov: *Foundation,* Avon Books, New York 1964.

Elisabeth Noelle-Neumann: *Die Schweigespirale* (The Spiral of Silence), R. Piper & Co., Munich 1980. I owe a number of the following quotations to this book.

Solomon E. Asch: "Group Forces in the Modification and Distortion of Judgments," in: *Social Psychology,* Prentice-Hall Inc., New York 1952, p. 452.

E. Noelle-Neumann's book contains many of her own investigations of these problems with the aid of demoscopic methods.

Jean-Jacques Rousseau: *Dépêches de Venise, XCI* (Dispatches from Venice). La Pléiade. Gallimard, Paris 1964, Vol. 3, p. 1184.

James Madison, in: The Federalist, 1788, No. 49, February 2.

Alexis de Tocqueville: *Autorité et Liberté* (Authority and Freedom), Rascher, Zurich, Leipzig 1935, p. 55.

James Bryce: *The American Commonwealth,* Macmillan, London, Vol. II, Part IV, Chapter LXXXV, p. 337.

Guy de Maupassant: *Bel Ami.*

Walter Lippmann: *Public Opinion,* The Macmillan Co., New York 1922, 1954.

Niklas Luhmann, in: *Politische Vierteljahresschrift,* Vol. 11, 1970, No. 1, pp. 2-28.

David Hume: *Essays Moral, Political and Literary,* Oxford University Press, London 1963, p. 29.

Jean-Jacques Rousseau: *Essais sur la Critique de la Culture* (Essays on the Critique of Culture).

Phase transition analogies in revolutions: W. Weidlich, in: H. Haken (editor), *Synergetik* (Synergetics), Teubner, Stuttgart 1973.

A. Wunderlin and H. Haken: "Lecture on the Project of Multilevel Analysis within the Framework of the Research Concentrating on Mathematization,"

University of Bielefeld 1980; see also H. Haken: *Synergetik. Eine Einfuehrung.* (Synergetics. An Introduction), loc. cit.

Ivan London: Preprint 1981. Alexis de Tocqueville: "L'Ancien Régime et la Revolution" (The Old Regime and the Revolution).

14. Do Hallucinations Prove Theories of Brain Function?

A good popular survey of neurology will be found in: "The Brain," Sci. Am. September 1979, 44; D. H. Hubel, T. H. Wiesel: The Journal of Physiology, *195,*
No. 2, 215, November 1968 (Experiments on the visual system of monkeys). J. D. Cowan, G. B. Ermentrout, in: Springer Series in Synergetics, Vol. 5, p. 122, loc. cit. (Theory of drug-induced hallucinations).

For epileptic seizures see: A. Babloyantz, in: Springer Series in Synergetics, Vol. 6, p. 180, loc. cit.

For Hebb's synapse see D. Hebb: *Organization of Behavior,* Wiley, New York 1979.

15. The Emancipation of the Computer: Hope or Nightmare?

For pattern recognition see: K. S. Fu: *Digital Pattern Recognition,* Springer, Berlin, Heidelberg, New York 1976; K. S. Fu: *Syntactic Pattern Recognition Applications,* Springer, Berlin, Heidelberg, New York 1976; in: Springer Series in Synergetics, Springer, Berlin 1979, Vol. 5, p. 176, loc. cit.

T. Kohonen: *Associate Memory—A System—A Theoretical Approach* Springer, Berlin, Heidelberg, New York 1978, and in Springer Series in Ergonomics, Springer, Berlin 1979, Vol. 5, p. 199, loc. cit.

Can computers be capricious? See J. Weizenbaum: *Computer Power and Human Reason,* W. H. Freeman & Co., San Francisco 1976.

16. The Dynamics of Scientific Perception; or, The Rivalry among Scientists

Thomas S. Kuhn: *The Structure of Scientific Revolution,* University of Chicago Press, Chicago 1970.

R. K. Merton: *The Sociology of Science,* University of Chicago Press, Chicago 1973.

Harriet Zuckerman: *Scientific Elite.* The Free Press, Macmillan Publishing Co., Inc., New York 1977.

René Thom: *Stabilitê Structurelle et Morphogênêse* (Structural Stability and Morphogenesis), Benjamin, New York 1972.

Tim Poston, Jan Stewart: *Catastrophe Theory and Its Applications,* Pitman Publishing Limited, London 1978.

E. C. Zeeman: *Catastrophe Theory. Selected Papers,* Addison-Wesley, Reading, Mass. 1977.

G. B. Kolata: Science *196,* 287, 350-351 (1977).

H. J. Sussmann and R. S. Zahler: Synthese *37,* 117-216 (1978); Behavioral Science *23,* 383-389 (1978).

P. Glansdorff, I. Prigogin: *Thermodynamic Theory of Structure, Stability, and Fluctuations,* Wiley, New York 1971.

G. Nicolis, I. Prigogin: *Self-organization in Equilibrium Systems,* Wiley Interscience, New York 1977.

R. Landauer: Phys. Rev. *A12,* 636 (1975).

Ronald Forrest Fox: Proc. Natl. Acad. Sci. USA, *77,* No. 7, 3763 (1980).

17. Summing Up

On the future of the universe see: F. J. Dyson: Rev. Mod. Phys. *51,* 447 (1979), also R. Breuer: *Vom Ende der Welt* (The End of the World), Bild der Wissenschaft, Stuttgart 1981, Vol. 18, No. 1, p. 46.

Kurt Goedel: "Ueber formal unentscheidbare Saetze der Principia Mathematica und verwandter Systeme" (Formally Indeterminable Theorems of the Principia Mathematica and Related Systems) I., Monatshefte fuer Mathematik und Physik, 38, 173-198 (1931); cf. also Douglas R. Hofstadter: *Goedel, Escher, Bach, an Eternal Golden Braid,* The Harvester Press Ltd., Hassocks 1979.

Index